Literature as a Lens for Climate Change

Literature as a Lens for Climate Change

Using Narratives to Prepare the Next Generation

Edited by
Rebecca L. Young

LEXINGTON BOOKS
Lanham • Boulder • New York • London

Published by Lexington Books
An imprint of The Rowman & Littlefield Publishing Group, Inc.
4501 Forbes Boulevard, Suite 200, Lanham, Maryland 20706
www.rowman.com

6 Tinworth Street, London SE11 5AL, United Kingdom

British Library Cataloguing in Publication Information Available

Library of Congress Cataloging-in-Publication Data

Names: Young, Rebecca L., editor.
Title: Literature as a lens for climate change : using narratives to prepare the next generation / edited by Rebecca L. Young.
Description: Lanham, Maryland : Lexington Books, 2022. | Includes bibliographical references and index. | Summary: "This collection offers practical approaches to using literature as a lens for teaching about climate change. Contributors share their classroom experiences and reflections to urge educators at all levels to prepare students for the challenges of a climate-changed world" —Provided by publisher.
Identifiers: LCCN 2021056694 (print) | LCCN 2021056695 (ebook) | ISBN 9781498594110 (cloth) | ISBN 9781498594134 (paperback) | ISBN 9781498594127 (epub)
Subjects: LCSH: Climatic changes in literature. | Climatic changes—Study and teaching | Environmental education. | Environmental responsibility—Study and teaching.
Classification: LCC PN56.C612 .L58 2022 (print) | LCC PN56.C612 (ebook) | DDC 809/.9336—dc23/eng/20211227
LC record available at https://lccn.loc.gov/2021056694
LC ebook record available at https://lccn.loc.gov/2021056695

For William and Elsie who must live in the world
we and our students will leave them.
And to all who have inspired and supported the contributors to this book—
especially our students.

Because stories really can make a difference.

Contents

Foreword

As I am writing this, a study in the journal *Nature Climate Change* raises new concerns about a potential collapse of the Atlantic Meridional Overturning Circulation (AMOC).[1] Published just a few days after this year's Earth Overshoot Day on July 29, and a few days before the release of the Sixth Assessment Report of the Intergovernmental Panel on Climate Change, the study finds that the AMOC, which is responsible for the temperate climate in much of North America and Europe, is not only slowing down but showing several signs of imminent collapse. If it does collapse, it could bring extreme cold to Europe and parts of North America, raise sea levels, and disrupt monsoon patterns.

Connoisseurs of popular culture will recognize this scenario from movies like Roland Emmerich's disaster sci-fi flick *The Day After Tomorrow* and science fiction novels like Kim Stanley Robinson's Science in the Capital trilogy. In both cases, the creators were aware of earlier scientific studies of the AMOC and used narrative techniques such as estrangement and extrapolation to give their audiences a sense of what it *might be like* to live in a future world where the current has collapsed. But rather than just dreaming up changed climatic conditions, Emmerich and Robinson put *people*—characters—into those conditions and invite viewers and readers to feel along with them, viscerally, as they live through the harrowing moment when environmental crisis turns into catastrophe. And because their works are fiction (and to varying degrees determined by economic concerns) they aren't strictly bound to scientific accuracy.

As someone who teaches environmental narratives regularly in university classrooms, I am aware of the wide range of reactions that young people have in response to them. Some of those responses I expect, and some are utterly surprising. Moreover, they seem to change from year to year, mirroring different levels of awareness and alarm, new scientific insights, personal affectedness, and media consumption, along with game-changing events such as a global pandemic (I taught my most recent class on climate fiction online due to COVID-19 regulations), and a myriad of other psychological, cultural, political, economic, and ecological factors. As Matthew Schneider-Mayerson has shown in a series of empirical studies, the reception of climate fiction is much more complex than often expected.[2] The classroom is no exception and yet, as all teachers know, it does make a difference whether a given text is read "for fun" or thoroughly analyzed and discussed in the context of a class. Aside from the fact that students might have never read an assigned text unprompted in their spare time, it is also the pedagogical concept and context that influences their reception in the classroom.

Rebecca Young is acutely aware of what cultural texts can do in an educational setting. Her 2018 book *Confronting Climate Crises through Education: Reading Our Way Forward* advocates positioning ecopedagogy as the central organizing principle of curriculum and assessment design. In this edited volume, she has collected essays from educators around the world who not only share her belief in the power of storytelling, but also carry both the belief and the storytelling into their own classrooms. In her introduction, Young reminds us that young people today are faced with multiple interlocking environmental crises that were set in motion long before they were born, and that in the future they will have no choice but try to fix. Comparing this daunting task with that faced by the young hobbits in Tolkien's *The Lord of the Rings*, Young argues that we must help today's youth to tackle it by making not only Fridays for future, but every day of the week. What this means in practical terms for educators is that they must ask themselves, on the one hand, how to contextualize scientific knowledge to make it personally and culturally relevant for their students and, on the other hand, how such relevance might resonate beyond the classroom and engender prosocial action. These are big and important questions that, once again, have no easy answers. Not only does personal and cultural relevance depend highly on context, but empirical research also suggests that relevance and awareness do not necessarily translate into changed behavior.[3] To this we must add another complicating factor, namely that it is extremely difficult to trace in a scientific study the myriad ways in which a classroom experience may impact a young mind, either immediately or many years later.

All the more important are the insights that teachers present in this volume based on their own classroom experiences. It is in part the diversity of the contributions, both in geographical terms and in terms of the type of institution and education level, that makes them valuable. Whether they reflect on the experience of teaching a young adult novel in a college course in Minnesota or on the experience of preparing preservice educators in Johannesburg,

South Africa, to teach secondary students about climate change, the challenge faced by all contributors—and by educators more generally—is to engage young minds in an issue that might be perceived by them as boring, threatening, or anything in between.

Young suggests that, if we accept this challenge, environmental narratives—both fiction and nonfiction—can facilitate classroom conversations about climate science and other ecological topics, and that such conversations can be an important entry point into new ways of thinking. This seems to have been the case for the classes taught by the volume's contributors, and I have made similar experiences in my own teaching. When I taught my ecofiction class online during a lockdown last semester, students told me in our final discussion that contemporary environmental issues now resonate with them in a new way. One of the texts we read in that class was T. C. Boyle's *A Friend of the Earth*, a speculative near-future novel that features visceral evocations of extreme weather, from scorching heat waves to powerful storms. As parts of Europe are inundated by record-breaking floods while other parts burn, I can't help but wonder whether my students now think back to Boyle's fictional rendering of California in 2025, which they all thought was probably too extreme but nevertheless found disturbing. One thing that environmental narratives can do for us in a classroom is indeed to get us talking about an issue. A related and equally important effect is that they can open up our imagination, allowing us not only to envision ecological futures but also to relate them to ourselves and to our present situation in what the science fiction scholar Tom Moylan once called an "enlightening triangulation."[4] In ecopedagogy, as in the rest of our lives, we urgently need such triangulations.

Alexa Weik von Mossner
University of Klagenfurt

NOTES

1. Niklas Boers, 2021, "Observation-based early-warning signals for a collapse of the Atlantic Meridional Overturning Circulation." *Nature Climate Change*. doi: 10.1038/s41558-021-01097-4.

2. Matthew Schneider-Mayerson. 2020. "'Just as in the book'? The influence of literature on readers' awareness of climate injustice and perception of climate migrants." *ISLE: Interdisciplinary Studies in Literature and Environment* 27 (2): 337–364; Matthew Schneider-Mayerson, Abel Gustafson, Anthony Leiserowitz, Matthew H. Goldberg, Seth A. Rosenthal, and Matthew Ballew. 2020. "Environmental literature as persuasion: An experimental test of the effects of reading climate fiction." *Environmental Communication*. Doi: 10.1080/17524032.2020.1814377.

3. Suzanne Keen. *Empathy and the Novel* (New York: Oxford University Press, 2007).

4. Tom Moylan. *Scraps of the Untainted Sky: Science Fiction, Utopia, Dystopia* (Boulder: Westview, 2000), xvi.

Acknowledgments

I am grateful to the educators who have been willing to share their excellent ideas and precious time with me in support of this project. I have enjoyed learning about their teaching experiences and collaborating in their efforts to bring climate literacy to every classroom, every student. Despite geographical distance, we are connected by the shared conviction that while education alone may not solve climate crises, it is certainly where understanding and action can begin.

Introduction

Rebecca L. Young

A classic among nations and across generations, Tolkien's *The Lord of the Rings* exposes a similar truth for today's audience as it did for its first readers: the innocents of the world often bear the weight of its injustices. Like Frodo, Sam, Merry, and Pippin—the young hobbits of the Shire who save all of Middle Earth—today's youth carry an equally unfair, devastating burden:

> No one answered. The noon-bell rang. Still no one spoke. Frodo glanced at all the faces, but they were not turned to him. All the Council sat with downcast eyes, as if deep in thought. A great dread fell on him, as if he was awaiting the pronouncement of some doom that he had long foreseen and vainly hoped might after all never be spoken. . . . At last with an effort he spoke, and wondered to hear his own words, as if some other will was using his small voice.
>
> "I will take the Ring," he said, "though I do not know the way."[1]

As Frodo sits among the secret council of elves, dwarves, and men to decide what should be done with the One Ring—that which holds the power to destroy—his quiet recognition of the formidable task ahead shapes the fate of Middle Earth. Today's young people have accepted a similar challenge. Theirs too is an unfathomable enemy, one that thrives on exploiting the human lust for wealth, power, dominance but that persists on the more benign desires of everyday comforts, of personal indulgences. They too face the seemingly impossible task of overcoming a force set in motion through no fault of their own. We should not look to them. The climate crisis is not theirs alone to confront, but it will be their fight that determines how much else is lost and what may yet be saved.

The antagonist of *The Lord of the Rings*, Sauron's evil—embodied in the form of unnatural creatures and the destruction of nature—ravages Middle Earth as its inhabitants turn inward to their own affairs or battle each other in conflicts over land and power. This threat can no longer be ignored when Treebeard, carrying hobbits Merry and Pippin, walks into a decimated forest, its trees now fueling the creation of the corrupted wizard Saruman's Orc army. Peter Jackson's film adaptation portrays this concrete visualization vividly as the camera pans vast devastation shadowed by an ominous horizon. Shocked, Treebeard laments, "Many of these trees were my friends, creatures I had known from nut and acorn!" His reflection that "a wizard should know better" is a turning point—a recognition that the threat is no longer looming but immediate. He calls the Ents to war, declaring "there is no curse in Elvish, Entish, or the tongues of men for this treachery."[2] It is a moment of hope to imagine this powerful symbol of ancient wisdom joining forces with elves, dwarves, and men to fight for the natural world. As with the Ents, it takes painful recognition of a shared fate for other representatives of Middle Earth to join the fellowship sworn to protect it.

Amid the many battles fought to preserve good in their world and push back the symbolic forces promising to corrupt it, Frodo and Sam fight to carry out their arduous task of destroying the One Ring: "'I don't like anything here at all,' said Frodo, 'step or stone, breath or bone. Earth, air and water all seem accursed. But so our path is laid.'" Much like the path of today's youth, each of the four hobbits confront obstacles on their journey that serve as fierce reminders of what is at stake. They must also face their own doubts about what they will be able to accomplish against the destruction that surrounds them. As the wise companion Sam reminds Frodo about the fate that befalls them,

> We shouldn't be here at all, if we'd known more about it before we started. . . . But that's not the way of it with the tales that really mattered, or the ones that stay in the mind. Folk seem to have been just landed in them, usually—their paths were laid that way, as you put it. But I expect they had lots of chances, like us, of turning back, only they didn't.[3]

Fortunately for all those battling with them, the young hobbits do not abandon their task in retreat just as the Elves renew their alliance to men and vow to fight alongside them and the dwarves. When survival depends on protecting your home and all its inhabitants, there can be only one response: unite. It is only when the realms of Middle Earth come together in its defense that there is hope of victory—hope for survival.

While Tolkien's message cautioned against unbridled industrialization, it resonates as poignantly in the climate crisis we face today—perhaps more so.

Similar recognitions of how division and strife distract recently played out on our screens in the popular *Game of Thrones* series adapted from George R. R. Martin's novels. Rapt audiences tuned in as imaginative metaphors for our own vice-driven dramas unfolded in epic battles among Westeros' seven kingdoms as characters were scheming for wealth, competing for resources, grasping for power. So immersed are they in defending their own politics, identities, and freedom that they fail to recognize the enemy that threatens to destroy them.

Exasperated by his efforts to convey the magnitude of the situation, protagonist Jon Snow asks the cynical but contemplative Tyrion Lannister, "How do I convince people who don't know me that an enemy they don't believe in is coming to kill them all?" Tyrion's response is a fitting reflection of the dilemma climate scientists face in convincing the public to take seriously data regarding climate change: "People's minds aren't made for problems that large. White Walkers, the Night King, Army of the Dead. It's almost a relief to confront a comfortable, familiar monster like my sister."[4] And so we too divert our attentions into problems more easily grasped—wasting time on these distractions as we try to pretend that no globally more ominous confrontation is already advancing against us. As environmentalist writer Wendell Berry cautions of our own gravely serious game of thrones, "Whether we and our politicians know it or not, Nature is party to all our deals and decisions, and she has more votes, a longer memory, and a sterner sense of justice than we do."[5]

When the only option for survival in this fantasy eventually becomes clear—that the kingdoms must put aside the differences that pit them against one another and confront the Night King's army together—audiences may be reminded of the lesson that saves Middle Earth. Victory in both imaginative worlds is hard won as many are lost in the final defining battle of each. But the characters' ability to unite—despite all the differences that divide them—is the resolution that we celebrate. As we share their fictional triumphs, we doom ourselves to heartbreak if we fail to realize our parallel.

Despite their flaws, which are a cautionary mirror of our own, we root for the characters who defend humanity and all who live in the natural world. We believe they will win. We need them to win.

Young people today are as blameless of climate change destruction as Tolkien's hobbits were of the forging of the ring of power. Fortunately for us, they also share the hobbits' courage. Just as the burdened Frodo confesses, "I wish it need not have happened in my time," today's youth are aware of the injustice that has been foisted upon them. Consequences of previous generations' failure to act are already in motion with impacts expected to worsen significantly in this century. As adults, we owe today's children at least the grace of this acknowledgment. As Gandalf responds to Frodo's frustration, yes, we all regret living to see climate change being allowed to accelerate as it has. That said, our responsibility remains to decide "what to do with the time that is given us."[6]

And yet, shockingly, "We are still educating the young as if there were no planetary emergency."[7] As environmental educator David W. Orr cautions in the preface to his more recent *Dangerous Years: Climate Change, the Long Emergency, and the Way Forward*, "There can be no sustainable and just economy and harmony between humans and natural systems in a society ruled by fear, threats, violence and war. A house so divided will fall." He notes that because of our divisions,

> the necessary changes will most likely begin in neighborhoods, cities, states, regions, and networks of global citizens. They must start at a manageable and comprehensible scale by a process of trial and error. And they must cascade up to change the larger systems of governance and economy.[8]

Driven in part by Greta Thunberg's call to "act as if our house is on fire,"[9] the school strikes that energized youth prior to the COVID-19 pandemic were beginning to shape one such global network. Working to galvanize efforts around climate action by striking from school and organizing awareness campaigns, students worldwide were becoming a powerful force for climate education. As the media reported on protests, we also saw more climate-related news coverage in general—interviews with climate scientists and environmental activists, commentaries on the connection between natural disasters and global warming, political debates about how best to manage costs and mitigate impacts of climate disruption. Notable on the list of demands that emerged from this momentum driven by young people is "compulsory comprehensive education on climate change and its impacts."[10] This speaks broadly to the crux of the crisis: we desperately need a global citizenry that understands the complex interrelated causes and consequences of a warming planet. As Wendell Berry has noted,

> Once our personal connection to what is wrong becomes clear, then we have to choose: we can go on as before, recognizing our dishonesty and living with it the best we can, or we can begin the effort to change the way we think and live.[11]

Armed with this knowledge, students may have a shot at uniting their generation in action. They may even succeed at persuading the rest of us to join forces with them before it is too late.

Climate change is the formidable enemy of our time, our world. Recognizing that the chance we still have to confront it depends on swift, unprecedented intervention, this collection of chapters is offered in support of our best defense: educators and students.

Literature in all its forms can serve as a lens to shape the kind of public education youth are asking for by helping educators answer two important questions:[12]

1. How do we provide context to make scientific data feel personally and culturally relevant?
2. To what extent can we promote a learning environment that transforms apathy and hopelessness into engaged, prosocial action?

As a universal resource, environmental narratives can facilitate conversations about climate science in classrooms around the world. Fiction and nonfiction that address the implications of a changing climate can foster student inquiry that leads to empathy. Stories that encourage exploration into patterns of human behavior that contribute to climate change, or cause us to deny or ignore it, can focus knowledge toward recognition. With support, recognition can lead to informed action.

Cognitive cultural studies illuminate the universal art of storytelling in humanity's evolutionary history as a means for making meaning out of information and for communicating that meaning within social groups. Neuroscientific investigations of Theory of Mind (ToM) help demonstrate why the sharing of stories has been beneficial across cultures, suggesting that narratives may elicit or potentially deepen empathy, which in turn might help guide future decisions. As a concrete reference point for ToM as it relates to reading, Lisa Zunshine offers this explanation in *Why We Read Fiction: Theory of Mind and the Novel*:

> By imagining the hidden mental states of fictional characters, by following the readily available representations of such states throughout the narrative, and by comparing our interpretation of what the given character must be feeling at a given moment with what we assume could be the author's own interpretation, we deliver a rich simulation to the cognitive adaptations constituting our Theory of Mind.[13]

She further explains that it matters little whether our interpretations of characters' thoughts are correct because it is the act of simulation itself that we enjoy, a form of practice for better understanding our own experiences with people whose intentions we may correctly or incorrectly intuit. As professor of English Patrick Colm Hogan notes, one of the reasons we "respond emotionally to literature" even though "we know perfectly well that the characters are not real" is that "we mentally simulate the experiences of a character from his or her point of view."[14] The implicit conclusion that we can learn from these simulations suggests that the experience may have transformative effects in our real lives.

When mild-mannered, people-pleasing Walter in Jonathan Franzen's *Freedom* screams at his audience that "WE ARE A CANCER ON THE PLANET!,"[15] readers feel his heartbreaking rage, experiencing viscerally the buildup of emotion that causes this break in character. As Walter's anger and the novel's title devastatingly illuminate, our lifestyles—our rigorous, nearly vicious protection of individual freedoms—have led to the detriment of the environment and "EVERY OTHER SPECIES IN IT!"

Walter's outburst may not be a call to action, per se, but it prompts his own redemptive lifestyle change and leaves readers shaken. A glaring confrontation lies within the pages of *Freedom*: we see ourselves, our family, our neighbors, our country, our world exposed. If we have been paying attention, we too are kept "awake at night" by Walter's fear that "all the real things, the authentic things, the honest things are dying off."[16] No one wants to be the cancer.

Readers who can empathize with Walter's character surely find themselves questioning the hidden costs behind their own perceived freedoms. Like Walter, they may be motivated by this reflection to call it out for others, perhaps even emboldened to make a change. In this simulation lies the transformational potential of literature, which contributors to this volume explore from a variety of perspectives as they share their approaches to teaching about the environment and climate change.

The goal of studying narratives that tell a particular climate story, especially those that help explain the phenomenon behind a specific impact, is to meaningfully contextualize information for students unfamiliar or uncomfortable with the science or experience. As Alexa Weik von Mossner notes in *Affective Ecologies: Empathy, Emotion, and Environmental Narrative*, narrative

> is a means for making sense of the world; not only of the imaginary world on the pages of a book or on the silver screen of a movie theatre, but also of the actual world in which we live out our lives.[17]

Narratives can help bridge the knowledge gap that exists between climate science and one's personal understanding (or misunderstanding) of it because

> [climate change] is a phenomenon that is too abstract and too vast in its spatial and temporal dimensions to allow for easy visualization. It exists in scientific graphs and numbers and is encapsulated in iconic images such as receding glaciers, devastating storms, and drowning polar bears.[18]

Monstrous representations of climate change that are too overwhelming to confront.

F. Stuart Chapin III frames it this way:

> As scientists, we are trained to avoid speaking in ways that touch people's souls. New forms and venues of dialogue must emerge through words, images, and tones that enable society to see and hear more clearly the changes that are occurring, to feel deeply and personally their importance, and to recognize the connections between our personal and collective choices and the trajectory of life on this planet.[19]

Rachel Carson confronts this dilemma forthrightly in her opening to *Silent Spring*. The cautionary message of "A Fable for Tomorrow" proved an incredibly effective introduction to her groundbreaking research on DDT because it spoke so meaningfully about the disastrous implications of pesticide use: "No witchcraft, no enemy action had silenced the rebirth of new life in this stricken world. The people had done it themselves."[20] As Weik von Mossner notes,

> Carson's "fable" is only two pages long, but it provides what many more recent examples of popular science writing fail to provide—vivid sensory imagery that allows readers to simulate the described world in their minds and therefore experience it on the imaginary level.[21]

Carson was certainly not trying "to avoid speaking in ways that touch people's souls." While she was vilified and dismissed at the time, her work launched a movement.

Narratives have the capacity to help make these realities more concrete because they "offer people affective and experiential engagement with climate change on the imaginary level."[22] In a chapter titled "Alluring Visions: Hope, Desire, and the Affective Appeals of Ecotopia," Weik von Mossner analyzes this effect through the lens of Kim Stanley Robinson's *Pacific Edge*:

> While it seems to Kevin that he is "without a doubt the unhappiest person in the whole world," he at the same time finds solace in an environment that "pulse[s] green and amber, jumping with his heart, glossy, intense, vibrant, awake, alive" (K.S. Robinson 1988, 326.) Readers are cued to simulate that vibrant world in their minds and, by an implicit comparison, find their own world lacking. This is the kind of imaginative engagement that ecotopian novels and other literary texts can provide because, within certain boundaries, they give readers the freedom to envision what they find desirable.[23]

Specifically linking the act of reading literature to learning about a scientific phenomenon or environmental dilemma may help foster students' understanding of the science by engaging their ability as cognitivists to interpret and predict the motivations and behaviors of characters who are impacted by it. Through the relative safety of a literary lens (particularly a fictional one), students may glean significant connections between broader environmental concerns and specific human behavior patterns that contribute to them. They can use this contextual understanding to inform their responses to the real-world problems mirrored by the narrative, including how they work to collect and share information, how they envision and develop potential solutions, how they communicate or enact their ideas.

Suzanne Keen's work on narrative empathy helps illustrate why this approach to pedagogy may have meaningful results with respect to educating students about the environment. As Keen notes in *Empathy and the Novel*, "The very fictionality of novels predisposes readers to empathize with characters, since a fiction known to be 'made up' does not activate suspicion and wariness as an apparently 'real' appeal for assistance may do."[24] Thus, in the case of climate, science, and speculative fiction that present characters in climate-changed situations and conflicts, readers may be more willing to engage with specific concerns than they normally would be because the novels' presentations of them are less confrontational or threatening to their personally held beliefs, opinions, or misunderstandings about climate change as it relates to their own situations. In Keen's evaluation,

> Fictional worlds provide safe zones for readers' feeling empathy without experiencing a resultant demand on real-world action. This freedom from obligation paradoxically opens up the channels for both empathy and related moral affects such as sympathy, outrage, pity, righteous indignation, and (not to be underestimated) shared joy and satisfaction.[25]

While Keen and others emphasize that more research is required to understand exactly how fiction may affect readers' emotions and therefore, potentially, their real-world reactions to such emotions, she suggests that "narrative fiction may in fact enhance the potential for subjects to respond feelingly to situations and characters, disarming them of their customary suspicions and learned caution."[26] Given the urgent need for climate change education and action in all parts of the world, cultural narratives that serve as a stimulus for engaging with climate-related issues will help educators safely initiate or further explore conversations about concerns for environmental justice both locally and globally.

Importantly, these texts and the conversations they encourage are not meant to instigate negativity in the form of blame or despair. Instead, readings and discussions should act as a lens for processing information in a proactive manner. As Octavia Butler's character in *Parable of the Sower* advises his daughter, who is eager to frighten friends and neighbors into taking action for survival,

> It's better to teach people than to scare them, Lauren. If you scare them and nothing happens, they lose their fear, and you lose some of your authority with them. It's harder to scare them a second time, harder to teach them, harder to win back their trust. Best to begin by teaching.[27]

Despite the wisdom in Reverend Olamina's caution, today's readers may be more persuaded by Lauren's admonition to her friend, that "we can get ready. That's what we've got to do now. Get ready for what's going to happen, get ready to survive it, get ready to make a life afterward."[28] Butler's/Lauren's 1993 caution could not have proved more correct: "People have changed the climate of the world. Now they're waiting for the old days to come back."[29] Fortunately, young people seem to realize that this is not an option. They want to be prepared to lead.

Through climate education we can support readiness for what is to come as well for what can be done. Interdisciplinary scholar Nancy Easterlin highlights the impact of literature as a pedagogical lens, connecting it to the ways we learn:

> Art is, in effect, a way of knowing and coping with the world, one that initially, perhaps, served to strengthen human groups. While its function has mutated somewhat in modern culture, art still serves as a way of coping and knowing—of exercising problem-solving skills, of imagining alternatives, of simply taking a break from the strains of daily living.[30]

As a resource to help people make meaning out of information related to climate change, the art of aboriginal storytelling in Australia has an interesting history. In *The Water Will Come: Rising Seas, Sinking Cities, and the Remaking of the Civilized World*, Jeff Goodell explores the role of myth in native Australians' preservation of geological records through cultural memory. Considering a hypothesis that "a ten-thousand-year old event such as sea-level rise could be the basis for aboriginal myths," Goodell cites this observation noted in a book about the dying Australian language Yidin:

> It is, however, worth noting that a theme running through all the coastal Yidinji myths is that the coastline was once where the barrier reef now stands (as in fact it was some 10,000 years ago), but the sea then rose and the shore retreated to its current position.[31]

The implications of this observation—"How could a story be told accurately, over and over again, for ten thousand years?"—became the basis for Australian linguist Nicholas Reid's investigation into aboriginal myths that reference sea levels.

As Goodell notes, the stories may "tell us nothing about what these aboriginal tribes thought or felt about the seas rising around them. But they do capture how deeply significant and strange this experience must have been, how inexplicable."[32] In an example he shares, we see the role of mythologies in helping humans make meaning out of observed changes:

> In the beginning, as far back as we remember, our home islands were not islands at all as they are today. . . . Then Garnguur, the seagull woman, took her raft and dragged it back and forth across the neck of the peninsula, letting the sea pour in and making our homes into islands.[33]

Storytelling to explain the phenomenon of sea-level rise offered tribes a way to preserve important information for future generations. According to Reid the existence of these stories reveals that rather than sea-level rise being a negligible change over "an individual's lifetime," the changing coastline would have been of great concern:

> Australia must in fact have been abuzz with news about this. There must have been constant inland movement, reestablishing relationships with the country, negotiating with inland neighbors about encroaching onto their territory. There would have been massive ramifications of this.[34]

These fascinating findings could serve as an excellent springboard for studying contemporary stories that document the sea-level rise we are experiencing today, including its implications for current and future generations. Oral traditions are but one example of communicating environmental change over time; there are many different forms of cultural storytelling that can similarly provide guidance on the presence of a change that may be difficult to comprehend from a human's short life span.

Young storytellers in the climate movement are doing just that. Eunice Andrada's "Pacific Salt" spoken word poetry performance is a call to action in its chronicling of the devastating impacts of sea-level rise on island nations, the opening a confrontation to those who ignore it: "Perhaps the further you are, the smaller we become."[35] The images she evokes are meant to inspire change but they are also a documentation of crises,[36] a record of abuses against a planet which inflict consequences upon people who are the least culpable.

The U.S. National Youth Poet Laureate Amanda Gorman offered the poem "Earthrise" at a Climate Reality training in 2018. Echoing the lesson explored in Butler's *Parable of the Sower*, Gorman advises her audience to approach information about climate change with courageous resolve rather than fear. Much as the fantasy worlds created by Tolkien and Martin do, this young poet dares audiences to "to dream a different reality, / Where despite disparities / We all care to protect this world."[37]

Stories about the environment heard from mentors have prompted other young activists to themselves become storytellers through film, music, and media hoping to spread awareness of the need for urgent action to their peers. Speaking with Al Gore during the TED Countdown Global Launch 2020, actor and musician Jaden Smith explained the motivation for his protective activism: "When I was about 11 years old, my surfing teacher taught me that the ocean was alive and my environmental teacher taught me that the ocean was dying," he explained. "From that point on, I've just had an undying commitment to the crisis—and I really feel the urgency to act now."[38] In the 2020 documentary *Brave Blue World*,[39] Smith and Water.org cofounder Matt Damon share hope that the spirit of ingenuity will guide us. Narrated by Liam Neeson, the film tells the inspiring stories of real-life innovators who are working to solve the world's water crisis.

Youth director for the conservation organization Youth Guardians, Xiuhtezcatl Martinez, uses a different medium to convey his climate message and inspire youth to join his efforts:

> Music is both a tool for resistance, and a medium to tell my story. My dad taught me that all life is sacred. When I was a little boy, we would always talk about our responsibility to protect our land, our culture, our earth as indigenous people.[40]

He has turned these lessons into activism. In addition to his music, Martinez offers a practical guide to getting involved in the climate fight in *We Rise: The Earth Guardians Guide to Building a Movement that Restores the Planet*. Imploring young people to join forces because they and future generations have the most to lose, he urges in his conclusion to the book, "The time has come for us to put aside everything that divides us and rise together like the oceans to turn the tides."[41]

Greta Thunberg uses the power of media to frame young people's perspective and situation quite directly. In her speech at the UN Climate Action Summit, she calls out politicians' "fairy tales of eternal economic growth" that delude the public, making plain the impact of the story adults have crafted:

> You are failing us. But the young people are starting to understand your betrayal. The eyes of all future generations are upon you. And if you choose to

> fail us, I say: We will never forgive you. We will not let you get away with this. Right here, right now is where we draw the line. The world is waking up. And change is coming, whether you like it or not.[42]

Using science as her weapon, this young activist is arming a generation with knowledge and indignation.

Recognizing that action on climate change must take many forms in the efforts of all stakeholders, contributors to this collection share their approaches to preparing young people for an epic twenty-first-century battle. In this story, it is Earth we defend. "Best to begin by teaching."

The authors of the following chapters represent a range of backgrounds in education and come from several different parts of the world. Each approach is thus unique in its guidance for harnessing the power of literature to create change. Just as the climate strikes have been driven by young people of all ages, ideas are presented for each level of education. Texts used as exemplars represent a range of literature (both fiction and nonfiction) that can support ecopedagogical goals. While text examples are specific to readers' abilities, most strategies can be adapted for use with very young readers to preservice teachers and other college students. Similarly, the research and resources referenced in each chapter are meant to provide instructors in a range of content areas with the tools they need to offer climate education. This effort can and should take many forms to be effective. Readers can access additional resources provided for some of the chapters at the Rowman and Littlefield webpage for this book.

The first chapter, "'It Wasn't Us!': Teaching about Ecocide and the Systemic Causes of Climate Change," provides an informative rationale for teaching climate literacy at any age. Referencing strategies that engage students with questions of agency, responsibility, and planetary transformation, Marek C. Oziewicz offers guidance for educators to develop ecopedagogy around literary narratives—and do so in ways that address the systemic contexts necessary for understanding climate change and its potential solutions.

Presenting ecocritical analyses as models, Suhasini Vincent explains how literature can inform students' understanding of the interrelated causes and consequences of climate change, particularly in developing parts of the world. "Amitav Ghosh and Arundhati Roy on Climate Change: A Pedagogical Approach to Awakening Student Engagement in Ecocriticism" illustrates the importance of analyzing literature through the lens of postcolonial ecocriticism, providing guidance on how to engage students in this effort.

Focused on the young adult novels *Dry*, *The Marrow Thieves*, and *Not a Drop to Drink*, "Climate Crisis Confluence, History, and Social Justice: How Race, Place, Privilege, Past, and Present Flow Together in YA Literature" explains an approach to using climate-based literature circles that engage students in discussion about how the impacts of climate change are disproportionately experienced. Pairing fiction with nonfiction, Anna Bernstein and

Kaela Sweeney help students investigate the contexts that contribute to gross inequities of the climate crisis.

In "Starting Points for Student Inquiry into Our Relationship with the Environment," Ryan Skardal explains a secondary or post-secondary instructional framework that interrogates students' understanding of human versus nature conflicts portrayed in environmental narratives. Through inquiry and personal engagement with a variety of text concepts, students are guided through a research process that challenges them to redefine relationships with nature in a climate-changed world.

In "Foregrounding the Value of Ecocriticism in a South African University Context," David Robinson shares the experience of preparing preservice educators to teach secondary students about climate change, particularly as its consequences relate to local environmental concerns. With reference to ecocritical approaches to a variety of texts, ideas are adaptable and represent the ways we can connect storytelling to students' lived experiences.

A close analysis of monologue in *A Midsummer Night's Dream* is the focus of "These Are the Forgeries of Jealousy: Nature Out of Balance." Timothy J. Duggan and Natalie Valentín-Espiet examine how engaging theater techniques to study Titania's speech can help students conceptualize the imagery Shakespeare uses to explore how human behaviors and relationships impact the natural world.

"Raising Environmental Awareness and Language Acquisition through Haiku" explains the value of poetry as a learning tool for exploring students' connection to the environment while helping them practice and refine language skills. Offering an overview of the ways traditional Haiku writers incorporate environmental themes in their poetry and suggesting creative ways teachers can use these models to support students' understanding of climate change, this chapter includes easily adaptable activities to support preservice educators or K–12 students.

"Introducing Sustainability Topics with Ursula Le Guin's "The Ones Who Walk Away from Omelas" and Richard Powers's "The Seventh Event" analyzes the ways educators can useshort stories to empower students and incite social change. Two celebrated works of fiction challengereaders' perceptions of the social contract and human/environmental interactions. In light of climate change and its impact, this ecocritical reading guide examines the ways each story positions readers to engage with complex ethical and environmental dilemmas.

In "Developmental Bibliotherapy and Cli-Fi: Helping to Reframe Young People's Responses to Climate Change," Judith Wakeman explores how school librarians can use the messages in climate fiction written for young adults in school-based programs intended to help students engage with climate education and convert their climate anxiety into hope, agency, and action.

We hope *Literature as a Lens for Climate Change* will serve educators of all backgrounds in every part of the world who are bold enough to answer our students' call to action. In the words of Amanda Gorman at the 2021 U.S. presidential inauguration,

"We will not be turned around / or interrupted by intimidation . . ."[43]

NOTES

1. J.R.R. Tolkien, *The Fellowship of the Ring*: *Being the First Part of The Lord of the Rings* (Boston, MA: Houghton Mifflin Company, 1954), 284.

2. J. R. R. Tolkien, *The Lord of the Rings: The Two Towers*, "The Last March of the Ents," directed by Peter Jackson (New York: New Line Cinema, 2002), DVD.

3. J. R. R. Tolkien, *The Lord of the Rings: The Two Towers* (Boston, MA: Houghton Mifflin Company, 1954), 320–321.

4. *Game of Thrones*, Season 7, episode 3, "The Queen's Justice," directed by David Benioff, D. B. Weiss, George R. R. Martin, Vince Gerardis, Ralph Vicinanza, Guymon Casady, Carolyn Strauss, et al., aired July 30, 2017, HBO.

5. Wendell Berry, endorsement statement for Charles E. Little's *The Dying of the Trees: The Pandemic in America's Forests*, 1995.

6. Tolkien, *The Lord of the Rings: The Fellowship of the Ring*, 60.

7. David W. Orr, *Earth in Mind: On Education, Environment, and the Human Prospect* (Washington, DC: First Island Press, 2004), 27.

8. David W. Orr, *Dangerous Years: Climate Change, the Long Emergency, and the Way Forward* (New Haven, CT: Yale University Press, 2016), xi-xii.

9. Greta Thunberg, "Greta Thunberg full speech at UN Climate Change COP24 Conference," December 15, 2018, *YouTube*, video, https://youtu.be/VFkQSGyeCWg.

10. Zoe Sayler, "Kids to Teachers: We Need to Talk about Climate Change," Grist, March 21, 2019, https://grist.org/article/kids-to-teachers-we-need-to-talk-about-climate-change.

11. Wendell Berry, qtd. in *Hope Beneath Our Feet: Restoring Our Place in the Natural World*, Martin Keogh, ed. (Berkeley, CA: North Atlantic Books, 2010), 32.

12. Rebecca Young, *Confronting Climate Crises through Education: Reading Our Way Forward* (Lanham: Lexington Books, 2018).

13. Lisa Zunshine, *Why We Read Fiction: Theory of Mind and the Novel* (Columbus, OH: The Ohio State University Press, 2006), 25.

14. Patrick Colm Hogan, "On Being Moved: Cognition and Emotion in Literature and Film," in *Introduction to Cognitive Cultural Studies*, Lisa Zunshine, ed. (Baltimore, MD: The Johns Hopkins University Press, 2010), 245.

15. Jonathan Franzen, *Freedom* (New York: Farrar, Straus and Giroux, 2010), 484.

16. Franzen, *Freedom*, 218.

17. Alexa Weik von Mossner, *Affective Ecologies: Empathy, Emotion, and Environmental Narrative* (Columbus, OH: The Ohio State University Press, 2017), 7.
18. Weik von Mossner, 139.
19. F. Stuart Chapin III, "The Planet is Shouting but Nobody Listens," in *Moral Ground: Ethical Action for A Planet In Peril*, Kathleen Dean Moore and Michael P. Nelson, eds. (San Antonio: Trinity University Press, 2010), 77.
20. Linda Lear and Rachel Carson. *Silent Spring* (UK: HMH Books, 2002), 3.
21. Weik von Mossner, 162.
22. Weik von Mossner, 141.
23. Weik von Mossner, 179.
24. Keen, Suzanne. *Empathy and the Novel* (New York: Oxford University Press, 2007), 4.
25. Keen, 4.
26. Keen, 13.
27. Octavia Butler, *Parable of the Sower*, reprint ed. (New York: Grand Central Publishing, 2019), 61.
28. Butler, 52.
29. Butler, 53–54.
30. Nancy Easterlin, "Cognitive Ecocriticism: Human Wayfinding, Sociality, and Literary Interpretation," in *Introduction to Cognitive Cultural Studies*, Lisa Zunshine, ed. (Baltimore, MD: The Johns Hopkins University Press, 2010), 263–264.
31. Jeff Goodell, *The Water Will Come: Rising Seas, Sinking Cities, and the Remaking of the Civilized World* (Little, Brown, and Company, 2017), 20.
32. Goodell, 23.
33. Goodell, 22.
34. Goodell, 23.
35. Eunice Andrada, "Pacific Salt," 2015, *YouTube*, Accessed March 12, 2021, video, https://www.youtube.com/watch?app=desktop&v=T4C2g-PHE4Q. See also her collection of poetry *Flood Damages*.
36. Andrada performed "Pacific Salt" at COP 15 in Paris after earning a Spoken Word for the World award.
37. Amanda Gorman, "Earthrise," 2018, *YouTube*, Accessed March 12, 2021, video, https://www.youtube.com/watch?v=xwOvBv8RLmo. See also the NAAEE website for lyrics: https://naaee.org/eepro/blog/earthrise-poem-amanda-gorman, 2019.
38. De Elizabeth, "Jaden Smith Speaks about the Climate Crisis at TED Countdown Global Launch 2020," *Teen Vogue*, October 10th, 2020, https://www.teenvogue.com/story/jaden-smith-climate-crisis-ted-countdown-global-launch-2020.
39. *Brave Blue World: Racing to Solve Our Water Crisis*, directed by Tim Neaves and Alexander Whittle. *Netflix*, 2020. https://www.braveblue.world/.
40. Gigi Buddie, "Introducing Our New Series on Indigenous Voices," *The Center for Sustainable Practice in the Arts*, September 18, 2020. https://sustainablepractice.org/2020/09/18/introducing-our-new-series-on-indigenous-voices/.
41. Xiuhtezcatl Martinez, *We Rise: The Earth Guardians Guide to Building a Movement That Restores the Planet* (Harmony/Rodale, 2017).

42. Greta Thunberg, "Transcript: Greta Thunberg's speech at the U.N. Climate Action Summit," September 23, 2019, https://www.npr.org/2019/09/23/763452863/transcript-greta-thunbergs-speech-at-the-u-n-climate-action-summit.

43. Amanda Gorman, "The Hill We Climb," January 20, 2021, https://www.cnn.com/2021/01/20/politics/amanda-gorman-inaugural-poem-transcript/index.html.

REFERENCES

Andrada, Eunice. "Pacific Salt." *YouTube*. 2015. Accessed March 12, 2021, video, 3:02. https://www.youtube.com/watch?app=desktop&v=T4C2g-PHE4Q.

Benioff, David, B. Weiss, George R. R. Martin, writers. *Game of Thrones*, Season 7, episode 3, "The Queen's Justice." Directed by Mark Mylod. aired July 30, 2017, HBO.

Berry, Wendell. Quoted in *Hope Beneath Our Feet: Restoring Our Place in the Natural World*. Edited by Martin Keogh, Berkeley: North Atlantic Books, 2010.

Buddie, Gigi. "Introducing Our New Series on Indigenous Voices," *The Center for Sustainable Practice in the Arts*. September 18, 2020. https://sustainablepractice.org/2020/09/18/introducing-our-new-series-on-indigenous-voices/.

Butler, Octavia. *Parable of the Sower*, reprint ed. New York: Grand Central Publishing, 2019.

Chapin III, F. Stuart. "The Planet is Shouting but Nobody Listens." In *Moral Ground: Ethical Action for a Planet in Peril*. Edited by Kathleen Dean Moore and Michael P. Nelson. San Antonio: Trinity University Press, 2010.

Easterlin, Nancy. "Cognitive Ecocriticism: Human Wayfinding, Sociality, and Literary Interpretation." In *Introduction to Cognitive Cultural Studies*. Edited by Lisa Zunshine. Baltimore, MD: The Johns Hopkins University Press, 2010.

Elizabeth, De. "Jaden Smith Speaks about the Climate Crisis at TED Countdown Global Launch 2020." *Teen Vogue*. October 10, 2020. https://www.teenvogue.com/story/jaden-smith-climate-crisis-ted-countdown-global-launch-2020.

Franzen, Jonathan. *Freedom*. New York: Farrar, Straus and Giroux, 2010.

Goodell, Jeff. *The Water Will Come: Rising Seas, Sinking Cities, and the Remaking of the Civilized World*. Little, Brown, and Company, 2017.

Gorman, Amanda. "Earthrise." *YouTube*. 2018. Accessed March 12, 2021, video, 4:29. https://www.youtube.com/watch?v=xwOvBv8RLmo.

Gorman, Amanda. "The Hill We Climb." January 20, 2021. https://www.cnn.com/2021/01/20/politics/amanda-gorman-inaugural-poem-transcript/index.html.

Hogan, Patrick Colm. "On Being Moved: Cognition and Emotion in Literature and Film." In *Introduction to Cognitive Cultural Studies*, edited by Lisa Zunshine. Baltimore, MD: The Johns Hopkins University Press, 2010.

Keen, Suzanne. *Empathy and the Novel*. New York: Oxford University Press, 2007.

Lear, Linda and Rachel Carson. *Silent Spring*. New York, NY: Houghton Mifflin Books, 2002.

Martinez, Xiuhtezcatl. *We Rise: The Earth Guardians Guide to Building a Movement That Restores the Planet*. Harmony/Rodale: 2017.

Neaves, Tim and Alexander Whittle, dirs. *Brave Blue World: Racing to Solve Our Water Crisis*, *Netflix*, 2020. https://www.braveblue.world/.

Orr, David W. *Earth in Mind: On Education, Environment, and the Human Prospect.* Washington, DC: First Island Press, 2004.

Orr, David W. *Dangerous Years: Climate Change, the Long Emergency, and the Way Forward.* New Haven, CT: Yale University Press, 2016.

Sayler, Zoe. "Kids to Teachers: We Need to Talk about Climate Change." Grist. March 21, 2019. https://grist.org/article/kids-to-teachers-we-need-to-talk-about-climate-change/.

Thunberg, Greta. "Greta Thunberg full speech at UN Climate Change COP24 Conference," December 15, 2018, *YouTube*, video, https://youtu.be/VFkQSGyeCWg.

Thunberg, Greta. "Transcript: Greta Thunberg's speech at the U.N. Climate Action Summit," September 23, 2019. https://www.npr.org/2019/09/23/763452863/transcript-greta-thunbergs-speech-at-the-u-n-climate-action-summit.

Tolkien, J.R.R. *The Fellowship of the Ring*: *Being the First Part of The Lord of the Rings.* Boston, MA: Houghton Mifflin Company, 1954.

Tolkien, J.R.R. *The Lord of the Rings: The Two Towers.* Directed by Peter Jackson. 2002; New York: New Line Cinema, 2002. DVD.

Tolkien, J.R.R. *The Lord of the Rings: The Two Towers*. Boston, MA: Houghton Mifflin Company, 1954.

Weik Von Mossner, Alexa. *Affective Ecologies: Empathy, Emotion, and Environmental Narrative.* Columbus, OH: The Ohio State University Press, 2017.

Wendell, Berry. Endorsement Statement. In *The Dying of the Trees: The Pandemic in America's Forests*, written by Charles E. Little, iv. New York: Penguin, 1995.

Young, Rebecca. *Confronting Climate Crises through Education: Reading Our Way Forward.* Lanham: Lexington Books, 2018.

Zunshine, Lisa. *Why We Read Fiction: Theory of Mind and the Novel.* Columbus, OH: The Ohio State University Press, 2006.

Chapter 1

"It Wasn't Us!"

Teaching about Ecocide and the Systemic Causes of Climate Change

Marek C. Oziewicz

In Geoff Rodkey's middle grade novel *We're Not from Here* (2019), a small group of human survivors seek asylum on the planet Choom. Their challenge is to convince the three local species—who, unfortunately, are familiar with humanity's environmental record—that humans are not a violent, destructive species. The Choom government gives human refugees a chance by allowing one family to stay on Choom as a test case. On his first day at school, the twelve-year-old protagonist Lan is bombarded with questions by his nonhuman classmates. All of them regard Lan as a representative of a monster species. Assuring everyone that he had never killed or eaten another human being, Lan insists that humans are peaceful. This is followed by a short exchange that leaves Lan speechless.

> "How can you be peaceful, when you destroyed your whole planet?"
>
> "I didn't—it wasn't us! It was just a few of bad people." There was a lump rising in my throat. I hadn't cried about Earth in ages, and I definitely didn't want to start now.
>
> "Why did you let them do it?"
>
> "We didn't—they just—they had powerful weapons, and—" I stopped in mid-sentence and took a deep breath, trying to hold back the tears.
>
> Half the hands in the room were up now, and they all reeked of fear. Not just fear. I smelled anger too.[1]

Eventually, Lan comes apart. "I've never killed anyone!" he sobs, "I've never *hurt* anyone. I'm peaceful. Humans are peaceful"[2] Emphasis in original. Although this exchange is part of a fictional narrative, it offers an excellent entry point to discussions about responsibility for climate change. Students are invited to consider that the questions Lan is asked are not about the past or future in space but about our present on this planet. In my teaching I have found that Lan's fumbling answers represent the most common form of helpless reasoning in response to climate anxiety young people feel today. The questions themselves make us uncomfortable. Are we really peaceful if we're destroying the planet? Or, perhaps, "we" are innocent for the destruction is orchestrated by a handful of "bad people"? If the latter, why are we unable to stop them? And who is the "we" majority and the evil few "them"? Alternatively, if all humans are implicated in the current ecocide, how do we determine the degree of responsibility borne by individuals, countries, and corporate players? And once we have a better sense of our individual and collective contribution to the climate crisis, how can we apply this knowledge to stand up for the planet in ways that make a difference?

Rather than offering answers, this chapter suggests that creating opportunities for our students to grapple with questions of responsibility and agency in story-rich contexts is crucial for the development of climate literacy. The focus is on conversations about the systemic causes of climate change. In my experience, students find these more difficult to grasp than direct cause-effect correlations like, say, that between the rise of carbon emissions and the warming of the earth's atmosphere. The chapter opens with a few general principles I developed for engaging my students in critical conversations about climate literacy. A reflection on student reactions to the notions of ecocide and planetary transformation is brought up to suggest that while young people today are often unaware of the systemic drivers of climate change, they are able to identify the destruction driven by neoliberal capitalism. The chapter then moves to consider strategies for introducing informed discussions about the ecocidal operations of the Capitalocene and the disproportionate responsibility of the world's rich for the climate catastrophe. Readings of Dr. Seuss's *The Lorax* (1971), Bill Peet's *The Wump World* (1970), and Shaun Tan's "Bears with Lawyers" (2018) are shown as helpful to introduce the concepts of producerism, extractivism, and fungibility. The chapter closes with an activity based on *We're Not from Here* (2019), in which students are invited to share opinions to questions about what the book leaves unsaid. The activities described have all been used with my students—the majority being teacher candidates or experienced teachers pursuing graduate degrees—in courses on children's literature I have taught at the University of Minnesota, Twin Cities. Many of these activities can be adopted to literature-based units about the systemic causes of climate change taught in middle and high school settings.

TEACHING CLIMATE LITERACY

As a comparative literature scholar and educator specializing in young people's literatures with the focus on fantasy, speculative fiction, and multicultural literature, I remained climate illiterate for the better part of my life. Looking back, I find it appalling how absent climate conversations have been in my education all through graduate school: then in my teaching and research too, including dozens of conferences, and my work for professional organizations in the field of children's and Young Adult literature. Even today—three decades after the rise of ecocriticism, environmental humanities, and specialized subfields and organizations—my sense is that we are barely beginning the conversation about climate literacy that should have started decades ago. For example, the first and so far only professional organization in my field to issue a statement on climate change is NCTE. Their "Resolution on Literacy Teaching on Climate Change" was adopted only in March 2019, but even NCTE does not have a definition of climate literacy yet. The first ever special issue of a children's literature journal focused on climate change, coedited by Lara Saguisag and myself for *The Lion and the Unicorn*, came out only in 2021. A literature review my graduate assistant Nick Kleese completed in late 2020 is downright scary: out of the total of 5,726 articles published to date in the top nine children's literature journals, only forty-one mention climate change at all and only fourteen directly engage with the topic.[3] Put otherwise, 99.76% of articles in our field have paid zero attention to climate change. (Sadly, this has also been the case with leading literacy organizations, where 99.12% of publications and conference presentations of the past decade have likewise completely overlooked climate change.[4]) We have only one textbook for teaching about climate change through literature, Richard Beach, Jeff Share, and Allen Webb's 2017 *Teaching Climate Change to Adolescents: Reading, Writing, and Making a Difference*, and only one literature-focused study of the affordances of environmental pedagogy for climate change education, Rebecca L. Young's 2018 *Confronting Climate Crises through Education: Reading Our Way Forward.* Dozens of other textbooks and collections for teaching children's literature, excellent as they are in other ways, ignore climate change altogether or merely mention it in passing. For instance, if an alien visitor to Earth was given MLA's *Teaching Young Adult Literature* (2020) or the second edition *Keywords for Children's Literature* (2021), they would never guess that these books were written on a planet careening toward a catastrophic climate collapse. Or that they were published in a decisive decade in which we have only ten years for "rapid, far-reaching and unprecedented changes in all aspects of society" to contain global warming at 1.5 degree Celsius by the end of the century—a level that the foremost body of international experts consider the threshold below which technologically advanced human civilization can survive.[5] While some may argue that climate literacy does not belong in the disciplinary purview of children's literature scholars, I fear that at this point such compartmentalization is actively contributing to climate illiteracy. Having spent the past five years in intense research about climate change, my assessment is that the fields of education and children's literature studies are abysmally behind when it comes to waking up to the scale and urgency of the challenge climate change poses to our civilization. This realization is foundational to my pedagogical approach. I see myself as a member of a climate illiterate generation that has long failed to act on this issue. Consequently, I consider my climate pedagogy and self-education to be my core responsibility as a humanist, educator, scholar, and placental mammal.

There are five general principles I have arrived at while teaching climate change. If the term "principle" seems too prescriptive, think of these as rules of thumb for climate pedagogy. The list is not exhaustive. It reflects my classroom experience and conclusions drawn from my research so far.

Principle 1: since no issue is more urgent for the survival of humanity than climate change, teaching about climate change should be at the heart of our educational practice. In the language used in *This Is Not a Drill: The Extinction Rebellion Handbook*, climate catastrophe "can no longer be ignored, denied or go unanswered by any beings of sound rational mind, ethical conscience, moral concern or spiritual belief."[6] What this means to me is that in our capacity as teachers we should include discussions about climate change—centering them, if possible—in any subject areas we teach. Considering that our students, BIPOC and white alike, will certainly experience unprecedented climate change events in their lives, teaching them disciplinary knowledge disconnected from universal climate literacy amounts to abrogating our responsibility as educators and adults. Education that fails to center climate literacy is like rearranging deck chairs on the Titanic a day before the disaster. Since 2018 I have thus included a climate change unit and climate change themes in all courses I teach and supervise. Since 2021 I have also added the following "Declaration of Climate Emergency" in each of my syllabi:

> As your instructor and fellow Earthling, I feel it is my duty to inform you that our planet is currently under a multi-pronged assault resulting from the rapacious operations of neoliberal capitalism. This assault on the biosphere occurs in the name of short-term gain and corporate profit. It is abetted by inept institutions, including higher education, whose inaction on climate change is allowing ecocide to continue. I urge you to seek any and all peaceful means to rebel against the system that is driving extermination of life and wants to make you complicit in its crimes. Awareness is the power that liberates minds and makes

transformation possible. Whatever you learn in this course, I ask you to consider this knowledge as a tool to fight for radical climate action now and apply it in your own life. In anything you do, stand up for your Mother Planet.

Principle 2: While every effort should be made to raise students' climate literacy, in communicating the burning urgency for climate action teachers must consider that climate change is a traumatic issue. Since climate change is an existential threat, attempts to grapple with its specific aspects may give rise to eco-anxiety, climate guilt, grief, and hopelessness that cumulatively may lead to traumatic withdrawal. When this psychological defense mechanism kicks in, students may simply disengage, shutting out consideration and discussion of climate change altogether. For specific strategies about how to manage eco-anxiety in the classroom, teachers should consult a growing body of research on the topic—a field too large for this chapter to tackle.[7]

Principle 3: Although most issues of climate change are discipline-specific, addressing climate change is not primarily a scientific or technological challenge. It is a challenge for our imaginations and story systems. Vanishing habitats and species extinctions can be presented in numbers and charts, but the message becomes affectively far more compelling when told as a story like in Sir David Attenborough's Netflix documentary *A Life on Our Planet* (2020) or in Mordecai Gernstein's picturebook *The Boy and the Whale* (2017). Scholars in environmental humanities have long recognized that no amount of scientific evidence, on its own accord, is sufficient to "communicate—and hence trigger—the social and political changes needed to address climate change."[8] Psychological research, likewise, has shown that people tend to filter out scientific evidence to arrive at an understanding that supports their preconceived notions.[9] The public may also be insulated from truth by their religious or political ideologies or manipulated through misinformation campaigns by powerful vested interests.[10] Even for someone who accepts science, however, the hard data of climate change remains difficult to conceptualize.[11] Climate change happens to be what Timothy Morton calls a hyperobject: an entity so extremely complex and so "massively distributed in time and space relative to humans" that we are unable to wrap our minds around its totality or even specific aspects.[12] For example, can you really imagine the 51.8 gigatons of CO_2 the world emitted in 2018?[13] Or the three billion animals that perished in Australia's fires in 2020?[14] While these numbers matter, our minds are evolutionarily designed to process information in different scales. We think in stories and our cognitive architecture is hardwired for narrative understanding.[15] Counterintuitive as it may sound, literature, film, and art are not additional but the most important avenues for raising climate awareness and mobilizing climate action. Thirty years after scientific consensus has been reached about the nature and causes of climate change, it is not more science that we need. We need more stories: stories that envision the transformation of the ecocidal status quo and keep alive the hope that such a transformation is possible. Which brings me to my next point.

Principle 4: Teaching about climate change is a wasted opportunity unless it empowers students to rediscover hope for the planet and realize that they are part of a larger community of life. Over the past three decades, the dominant response to climate change in literature and film has been dystopian. My students have no problems whatsoever with naming books, films, shows, or games that project the collapse of our civilization. But they struggle to recall a single story that projects a hopeful planetary future. Falling victim to availability heuristic, many assume that the dystopian future is certain. Most of my students are part of Generation Z, born between 1995 and 2012. They grew up surrounded by dystopian, postapocalyptic, and post-disaster narratives in a world hurtling toward apocalypse. During their lifetimes, between 1990 and 2020, global fossil fuel consumption grew by about 30%, atmospheric CO_2 increased more than it did in the entire life span of the human species until 1990, and the global average temperature of the planet increased by about .5 degree Celsius.[16] Raised in a culture addicted to accelerating these catastrophic developments, my students have internalized the belief that imagining a hopeful future is naive, whereas imagining a postapocalyptic hell is reasonable. What they don't realize is that the explosion of dystopia is a historical aberration of our story systems that began in the Reagan years, largely as a reaction to the expansion of corporate world order imposed by neoliberal capitalism.[17] What they don't realize is that the rhetoric of "no alternative" is the rhetoric of neoliberal capitalism[18]—a system that, indeed, sees no alternative to its peculiar idea of growth based on accelerating monetization, subjugation, and exploitation of the biosphere. In this sense, although dystopias emerged as warnings against the ecocidal status quo, they have largely come to articulate the core message of the system they supposedly challenge: yes, capitalism destroys the planet, but no, there is no alternative. This is not meant to suggest that dystopia has no place in climate change literacy; the point, rather, is that dystopias are not enough to bring about the change we need. Research shows that instead of mobilizing action, dystopias tend to elicit despair, helplessness, and anger that lead to "ignoring or avoiding the topic."[19] In their larger cultural work, dystopias have thus contributed to the erasure of hope. It is against this large context that I believe we need stories that mobilize hope—and not just any hope but specifically hope for the planet. Reading/creating such stories, I contend, is a form of resistance to the rhetoric that the earth is doomed. It expresses an epistemological orientation of not giving up on the planet that I have elsewhere called planetarianism.[20] A two-level phenomenon, planetarianism is a biocentric philosophical commitment to stand up for the planet—thus, a counternarrative to the dystopian discourse of neoliberalism that renders ecocide unavoidable. It is also applied hope articulated through stories: a form of hope-as-resistance. On this level, planetarianism enables us to design alternatives to ecocide through stories that imagine a non-ecocidal sociopolitical system, a disanthropocentrized planet, and a biocentric, multispecies future that is worth living for. What I believe our classrooms need is less dystopia and more planetarianism. We need literature that helps us imagine the futures we want rather than the futures we dread.

Principle 5: Teaching and learning about climate change requires new vocabulary, new concepts, new frameworks, and a radical transformation of how we imagine ourselves in relation to the biosphere. As we strive to communicate the truth about climate change and the urgency for climate action, we need to jump-start our imaginations and radically enrich our vocabulary. Words and concepts are the most powerful tools we have to articulate new understanding; they empower us to name, notice, and engage with the overwhelming issues we face. This generative notion informs a spate of recent works focused on the transformative power of words: Robert McFarlane's *The Lost Words* (2018), Glenn A. Albrecht's *Earth Emotions: New Words for a New World* (2019), Matthew Schneider-Mayerson and Brent Ryan Bellamy's *An Ecotopian Lexicon* (2019), and Ayana Elizabeth Johnson and Katharine K. Wilkinson's *All We Can Save: Truth, Courage, and Solutions for the Climate Crisis* (2020). In my experience, perhaps the darkest legacy of climate illiteracy is that we lack words to adequately grasp the phenomena unfolding in front of our eyes. Or, even if students have heard the terms, they are unsure of their meaning. For example, how many students are familiar with the notions of slow violence, the Anthropocene, environmental justice, carbon footprint, speciesism, eco-apartheid, or ecocide? How many teachers can tell the difference between environmental education and ecopedagogy, degrowthism and green growth, biocentrism and anthropocentrism, environmental conservation and preservation—or explain why the difference matters at all? Scholars, likewise, have used terms like environmental literature, ecofiction, ecoliterature, and cli-fi in ways that ignore the cultural situatedness of each term and thus fail to indicate similarities, differences, and affordances of each label. All these are failures of conceptual mapping—failures that support climate illiteracy. They weaken our understanding of specific processes and blur our grasp of how any one concept is related to others. One challenge I have thus taken up is to offer my students some clarity on the terms we use, even as I introduce them to arguments and discussions that support alternative framings. On the most basic level, for example, I always start by helping my students grasp the relation between climate change and global warming. While global warming is indeed "the mother of all issues,"[21] it refers only to the measurable rise in the global mean surface temperature of the earth caused by anthropogenic activity, especially the burning of fossil fuels.[22] Climate change is a wider, less precisely quantifiable term. The United Nations climate agency is called the Intergovernmental Panel on Climate Change—rather than panel on global warming, environmental crisis, or the Anthropocene—precisely because climate change is the best term we have for an open-ended range of consequences of global warming which affect all living systems on the planet through complex feedback loops with processes like pollution, desertification, deforestation, species extinction, soil erosion, ocean acidification, biodiversity loss, the expansion of human populations, resource depletion, and others. Since the impacts of this ever-growing cluster of entanglements are not limited to the environment, I want my students to know that climate change is not just about the environment or temperature. It has equally serious social, political, and economic consequences for human societies, all nonhuman life, and the planet as a whole.

While the five points listed above are merely a start, I believe they are part of the effort we each should make to help create universal climate literacy. And fast. Our classrooms are ground zero for this effort and we are running out of time. Fortunately, the need for climate education is being increasingly recognized. In spring 2019, an NPR/Ipsos poll found that over 80% of parents in the United States on both sides of the political divide want their children to learn about climate change and over 86% of teachers support the need for climate education.[23] In May 2020 the New Jersey Board of Education added learning about climate change to its seven standards, making New Jersey the first state in the United States to mandate climate education at every grade level.[24] In another push for climate education, in November 2020, former education secretaries John King, Arne Duncan, and the governor of New Jersey Christine Whitman issued an open letter to president-elect Joseph Biden, urging him to support climate education nationwide. Stressing "the considerable need to address climate change in America's public schools," the letter identifies educating fifty million children enrolled in public schools as "a critical opportunity toward climate solutions," building resilience, and preparing "children and youth to advance a more sustainable world."[25] All of these initiatives are part of the effort to nurture Americans' climate literacy and offer unique opportunities for engaging literature in this process.

I should also note that climate literacy—sometimes called "climate change literacy"—does not have a widely accepted definition yet. While it certainly overlaps with the concept of "climate science literacy" defined by the American Association for the Advancement of Science as "an understanding of your influence on climate and climate's influence on you and society,"[26] climate literacy is a more integrated and multidisciplinary cultural competence. Building on Hiser and Lynch,[27] I take climate literacy to be a lived, emotionally charged, radical hope-driven, and action-oriented embrace of our responsibility to stand up for our biospheric inheritance—for all of Earth's living systems that sustain us and are currently reeling under multipronged assault from anthropogenic climate change. We and our students are the two generations who can become climate literate and make a difference for the planet's future. Achieving universal climate literacy in 2050 or later would merely turn us into Dr. Seuss's Once-lers: regretful but unable to undo the damage. The definition of climate literacy I have thus used centers the present moment and the anthropogenic causes of climate change. I want my students to know that we must act now: that today's climate change is driven by human activity in general and the ecocidal operations of neoliberal capitalism in particular. The direct relationship between neoliberalism and climate destruction[28] is a fundamental component of climate literacy. This component is also most revolutionary for it has been consistently obscured—to the point of misrecognizability—by the neoliberal discourse in all mainstream climate negotiations, starting with the 1992 Rio Earth Summit. Even in the text of the 2015 Paris Agreement, Amitav Ghosh points out,

> There is not the slightest acknowledgment that something has gone wrong with our dominant paradigms; it contains no clause or article that could be interpreted as a critique of the practices that are known to have created the situation that the Agreement seeks to redress. The current paradigm of perpetual growth is enshrined at the core of the text.[29]

This paradigm is likewise at the heart of techno-fix pronouncements about climate change championed by billionaires such as Bill Gates, whose recent book *How to Avoid a Climate Disaster* (2021) casts economic growth as the "solution" we need. Nor has it been challenged at the COP26 in Glasgow, where the 500-plus team of fossil fuel lobbyists was larger than any delegation representing a single country.[30] Given this ongoing legacy of denial and misrepresentation, the definition of climate literacy I have used refers to an understanding of how neoliberal capitalism has brought the biosphere to the edge of collapse, and how we all, together, have the responsibility to find an alternative system that will work for all life.

"IT WAS JUST A FEW BAD PEOPLE!"

Teaching about the systemic causes of climate change is hard. First, the word "system" refers equally to large structures of relations that are designed by humans, like the financial system, or educational systems, and to structures that exist "naturally" in nature, like the ecosystem or the digestive system. The "systemic causes" may thus be taken to refer to anthropogenic causes, natural causes, or the combination of both. Another difficulty is that most key words we use in discussing the systemic causes of climate change—like economy, social system, capitalism, or neoliberalism—are hyperobjects too. They refer to entities so dispersed in time and space relative to our human scale that we can never fully describe them. For example, a school is part of the educational system but is not the system itself, nor is any bank or hedge fund or even Wall Street—all of them expressions of our economic system—the economic system itself. Morton names these elusive qualities of hyperobjects "nonlocality," that is, none of the system's manifestations can be pinned down as the system itself;[31] "interobjectivity," that is, the system is never itself but a mesh constituted by a plurality of different objects;[32] and "viscosity," meaning that hyperobjects like the economy or capitalism stick like oil to everything they touch, making it feel like they have been everywhere all along.[33] It is important to acknowledge these challenges. Yet it is even more important to recognize that becoming literate about the systemic causes of climate change does not require perfect knowledge. In fact, the disingenuous call for "more research" has been at the heart of climate denial and doubt-sowing funded by corporate interests for decades.[34] Today, any claims that climate change is—supposedly—insufficiently understood and requires more studies before action is taken amount to an active withholding of recognition that helps maintain the violent, ecocidal status quo. We don't need complete knowledge. Adequate knowledge is just fine.

This begs the question of what constitutes adequate knowledge about the systemic causes of climate change. Virtually all scholarship on climate change is unanimous that the key driver of climate change is our current economic mode through which we organize the planet's resources, labor, production, and distribution.[35] The most common shorthand term for this mode is capitalism—as used, for example, in Naomi Klein's *This Changes Everything: Capitalism vs the Climate* (2014)—but a more accurate label is neoliberalism: a global economic system "premised on unfettered corporate profiteering enforced by repressive and juridical state apparatuses"[36] that arose in the mid-1970s. Think of it as a rigged, high-stakes monopoly game, which everyone is forced to play—except with their real lives and livelihoods—to ultimately lose against banker-players who have unlimited access to the game's money and are able to change the game's rules to promote their own interests. Another metaphor I found my students can relate to is Kim Stanley Robinson's explanation of the neoliberal world economy as a Ponzi scheme. When the resources the world is able to replenish are used by about August each year, the resources for the rest of the year are stolen from the future.[37] As in, I'm taking your apple now, but you'll get the apple that will grow next year. The ever-growing costs of this stealing from the future are dumped on the three groups that have little or no political voice—the poor, the unborn, and the nonhuman. This kicking of the environmental costs into the future creates the attritional wake that Rob Nixon has famously called slow violence: "a violence that occurs gradually and out of sight, a violence of delayed destruction that is dispersed across time and space, an attritional violence that is typically not viewed as violence at all."[38] Irrespective of the level of granularity you want to introduce to your students, the bottom line is that "we" have created a socioeconomic system which promises, urges, and rewards infinite growth on a finite planet—a system that, effectively, wages a war of conquest on all life in the name of more for the few. The "we" here is clearly not all of humanity. It is the financial and political elites of the Global North—including "gargantuan, transnational corporations"[39]—which, since the mid-1970s, have advanced what Naomi Klein calls the "three policy pillars of this [neoliberal] era: . . . privatization of the public sphere, deregulation of the corporate sector, and lower corporate taxation, paid with the cuts to public spending."[40] This is how we have arrived in a new Gilded Age,[41] in which America's public infrastructure is crumbling,[42] student debt is astronomical,[43] and a billionaire President is able to pay less federal income tax than a single adult without children making under $18,500 a year.[44] In this sense, when students say "It wasn't us! It was just a few of bad people,"[45] they do have a point. They express a general sentiment that we live in a system we did not choose, a system whose logic of unsatiable accumulation and infinite growth is driving the demolition of the planet. But we need more than rage, and here is where education comes in.

To help my students articulate their questions, assumptions, and grasp of the relationship between neoliberal capitalism and climate change, I have found it productive to facilitate discussion through two introductory activities. These can be done in conjunction or separately, and I have applied these in all of my courses over the past four years.

The first activity is watching a 2012 TED talk "Ecocide, the 5th Crime Against Peace" by British visionary lawyer and environmental activist Polly Higgins. In this talk Higgins introduces the notion of ecocide—the crime of destroying the earth's ecosystems through corporate or state actions—arguing that ecocide should be added to the list of existing crimes against peace. Higgins, who passed away in 2019, campaigned for ecocide her entire life and proposed it as a new international atrocity crime to the United Nations in 2010. Using a paper sheet, Google form, or VideoAnt platform, students were asked to (1) formulate a general response to Higgins's proposal and (2) answer a specific question like "What would be accomplished if this proposal became law and what else would need to be done?" Or, in other versions of this activity, "Comment on two statements that you find most insightful, most challenging, or most problematic."

The other activity is a questionnaire—on paper or in Google forms—in response to the following hypothetical story I created:

> The Guests arrived in 2022 [I have always used the "immediate future" year to when the activity takes place]. It was like the Second Coming. Just not what anyone expected. Armed with evidence they had collected on humanity's reckless destruction of the biosphere, they came to salvage whatever life remained. Except for humans.
>
> Unable to eliminate or ignore them, the Earth's leaders begged for another chance. The Guests considered. A highly advanced species who have long acted as gardeners of life in other galaxies, they unexpectedly gave us 12 months. 12 months to reshape our global civilization: our political institutions, social organization, and dominant technologies. Unless we make it sustainable—environmentally, politically, and economically—in 12 months they will start draining the planet's water. They will also relocate all surviving life forms. Except for humans.

This story is then followed by five questions, each with a possibility of additional explanation. (1) "Would we be able to change our ways?," (2) "Would governments, banks and corporations be motivated to re-channel their resources into creating a sustainable global order?," (3) "Do you believe that we can save the planet and create a future for your grandchildren that's worth living for?," (4) "Are sustainable technologies, energy and progress possible?," and (5) "What do we need to do to make them a reality?." In questions 1 to 4, possible answers included "yes," "no," and "it's complicated," each followed by an optional prompt: "In one sentence, explain your choice above." In more recent editions of this questionnaire, I also asked students two additional questions: "How concerned are you about climate change and its present/future consequences?" coupled with "How informed do you consider yourself to be about climate change?"

Both the ecocide activity and the alien ultimatum questionnaire have been evolving since I first introduced them in 2016. Although the phrasing was tweaked, the core questions remained the same. By the end of 2020, I was able to collect seventy-eight responses to Higgins's talk and forty-nine responses to the alien ultimatum scenario. These numbers include only responses from students who gave their consent to my use of their answers in my research; their names below are replaced with fictional ones. Based on this input, I see compelling evidence for a paradoxical situation in which students have a better grasp of the systemic causes of climate change than they realize, yet they stumble when it comes to envisioning alternatives to the operations of the ecocidal system. The most common mistake they make is an attribution error, in which they blame people—or human nature specifically—for what they had earlier correctly identified as outcomes of the way our socioeconomic system is organized.

For example, pretty much all students agree with Higgins that "the Earth needs a good lawyer." Some argue that "the only way to create change in today's day and age is to make it a law. The economy and laws are all anyone cares about today, so that is the way to reach the people" (Maya 2020). Others point out that "ultimately we are in desperate need of a team of lawyers that can advocate . . . for the rights of the planet" (Jeremy 2019). They understand that "the goal of including ecocide on the list [of 'superlaws'] is to show that all life, and not just human life exclusively, should be valued . . . and protected" (Eli 2016). Many students support Higgins's notion to consider Earth as a living being and a legal person with their own inalienable rights: "If one views the Earth as a commodity, it is much easier to kill and destroy the Earth because we are viewing it as an object," one student explains.

> If you view the Earth as another living thing, however, it gives you the responsibility of taking care of the Earth and preserving the natural world for future generations. Therefore, the difference in the perspective of which you view the Earth has a great impact on how you treat it. (Laura 2016)

Students also welcome a recognition of basic rights to life for all nonhuman species.

> I appreciated how she includes both human and animal persons when creating this address. Many humans fail to recognize that we are not the only species on Earth and all other animal people deserve to live just as much as we do. (Camille 2019)

That said, students also recognize major roadblocks to introducing ecocide as part of Earth Law. Some mention ideological assumptions of human superiority that would be hard to challenge: "Growing up in America and into certain religions, we believe that we are the dominant species, that everything on Earth was made for us" (Annie 2016). Others discussed questions of effective legal implementation, stating that while ecocide is "a very insightful" and "certainly needed" law, "It would be incredibly hard to push a legal duty onto all countries that they are forced to help with the environment and are banned from destroying it in the way we are today" (Mark 2019). Other students took issue with the ways our society is organized around money, private ownership, and the—often destructive—freedoms these concepts imply. As one student wrote,

> I think that the problem lies in the idea of ownership of land. In our society, certain geographical areas are thought of as belonging to specific individuals or groups, even though there are many other beings that need and could claim this

> space—other animals, insects, and plants which have the right to these areas as well. (Sinead 2017)

Even as they identified formal, legal, and political obstacles, by far the most discussed challenge to the law of ecocide has been economic factors. In this group of responses, opinions are divided between seeing the capitalism as structurally ecocidal and seeing the primary culprit in human nature, for which capitalism is merely an extension of our destructiveness. In statements representing the first position, students argue

> that holding big corporations accountable for ecocide is the first thing we can do. The first thing we have to do is stop the problem. We can clean the earth and recycle all we want, but if pollution continues at its current rate, it won't matter. (Alexa 2019)

Students also note that governments' inaction on ecocide results from corporate pressures. "Often, governments are too focused on economic impacts and influence in order to make the necessary changes. We see this with numerous issues today, where the government refuses to act because it could supposedly weaken the economy" (Brenda 2019). In the words of another student,

> There are major influences on the government that tend to only focus/enhance on making profit and ever-growing capitalist structure. . . . These laws that Higgins suggests are crucial, and more important to enact than ever, but in order to do this I feel as if there must be a complete turn over in our governmental system itself. (Agnes 2019)

Linked to a recognition that "money creates power, therefore, economics tops environmental concerns" (Natalya 2020) and "change won't happen unless powerful, white, CEO America wants it to" (Mara 2019) are many comments about human nature and the worldview of people in power.

> While I, for one, love this statement and agree . . . that we, as the human race, should take responsibility for ecocide, I don't think many would agree with me. Selfish and irresponsible, many people refuse to open the door for the person behind them, let alone set laws in place for helping entire countries stay afloat (literally). If it would raise taxes or set any fines in place that would affect them directly, I think most people (especially Americans) would lean towards capitalism and protest against [introducing ecocide as law]. (Camille 2019)

"Ecocide should be a crime, but it won't because the human race is stubborn and loves their power too much" (Erica 2016). Or, in another version of this sentiment,

> People have unlimited wants and the earth has limited resources. This is known as scarcity. Unfortunately, everyone feels that they are the most deserving of the resources, and the earth is quickly stripped of its resources because it cannot stand up for itself. (Maggie 2017)

The notion that the market system works against the planet also comes up prominently in responses to the alien visit scenario. Over 90% of graduate students and close to 80% of undergraduates are skeptical as to whether we would be able to transform our economy toward sustainability that respects all life on the planet, even in the face of an alien ultimatum which effectively threatens to eliminate humanity. While a number of answers argue that one year is too short a time for any real transformation, the real issue students identify, again, is human nature and the nature of capitalism. The many versions of the argument that humans are selfish, unable to cooperate, and insatiably greedy repeat similar formulations I saw in the Ecocide video responses. As one undergraduate wrote,

> I do not think humans will be able to change. Being able to change and wanting to change are very different things. If the human species really wanted to change and make a big difference, they would have done so already. For decades, science has told humans that the Earth is going to fail us if we continue what we are doing, but that has not stopped humans at all. If anything, humans have been doing more damage for the past decades. I do not see humans changing within 12 months or ever. Humans are selfish beings that have created the habit of making themselves the top priority. (Jennifer 2020)

Likewise, answering the question "Would governments, banks and corporations be motivated to re-channel their resources into creating a sustainable global order?," about two-thirds of students are highly skeptical. "I think that institutions themselves might resist the changes necessary simply because of the way they are designed" (Michael 2019). "Even in the event of the end of the world, I think that the capitalist society that we have formed would still be looking for money and profit" (Sara 2020). While some students, hesitantly, suggest that corporations may be willing to change their operations "if the leaders' lives were in danger" (Lauren 2020), many more say that the change would either be superficial or that it would not happen at all "because sustainability can simply not exist in capitalism" (Lee 2020). Several students suggested another outcome too. "I think companies would invest in the creation of underground bunkers, water stockpiles and space travel to escape their fate. The resources to fix the planet would not be utilized for that purpose" (Karen 2018). In the words of another, "It is more likely that they would work to escape Earth and the Guests [i.e., aliens], using their monetary resources to build rockets and take resources to settle on Mars" (Jill 2019). These overwhelmingly negative perceptions of our current economic system which gives power to the few over the well-being of everyone else and the planet contrast sharply with the students' staggering YES—about 78% for graduate students and over 90% for undergrads—to the two other questions: "Do you believe that we can save the planet and create a future for your grandchildren that's worth living for?" and "Are sustainable technologies, energy and progress possible?"

There are many conversations you can have with your students based on these activities. One emerging theme is the exploration of hope: I have been struck by how hopeful young people are about the possibility of an ecological civilization despite their grim assessment of the ecocidal nature of our current system. The knowledge gap is about how to make the transition. Can capitalism be saved, or do we need a complete systemic transformation? Will transformed capitalism remain itself, or will we find ourselves in a de facto post-capitalism? When students are shown the sharp contrast between their assessment of the present and their beliefs in the future, a learning space is created for asking deeper questions about the meaning and sources of hope. Students are invited to ask themselves why they believe we are able to create a post-carbon world. In my experience, even those who had earlier asserted that the problem is human nature often end up admitting that—out of the two "impossibilities"—it would still be easier to change our socioeconomic system than to change human nature. In particular, students are able to find many more specific examples when civil disobedience, protests, and citizen-led grassroots initiatives were able to radically transform societies than examples when social change was the result of one person's or one group's change of heart. Hope, as these conversations help them realize, lies in the power of human collective action to expose, protest, and end abusive practices. This hope needs to be told in stories too. To help my students realize that this source of hope they just named is real, I have used excerpts from Rebecca Solnit's *Hope in the Dark* (2015) and Vandana Shiva's *Oneness vs the 1%* (2020). Both discuss hope as a form of awareness that recognizes the realities of our ecocidal present but an awareness that empowers us to envision different futures whose realizations demand action. Hope, Solnit writes, "is a gift you don't have to surrender."[46] It is the power so critically necessary for defying the oppressive status quo that the entire rhetorical apparatus of the ecocidal system in which we live is premised on discursive operations that undermine hope for change.

One of those operations is the naturalization of capitalism through claims that it is a system of relations and attitudes that essentially reflect human nature. Many of my students are victims of this view. The thing is, once you accept that the wholesale devastation of the planet that neoliberal capitalism brings is merely an expression of human nature, there is very little anyone can do to stop it. We naturally destroy everything—this view holds—and we'll destroy our planet too because that is who we are. In helping my students confront this view, I start with a recognition that the rhetorical move of attributing ecocide to peculiarities of human nature has powerful appeal. Among other things, it plays into the narrative of sinful human nature at the core of many religious traditions; it also, implicitly, absolves us of any responsibility to act. After all, if destruction is our nature, we have no more power to control it than we have controlling our livers. However, this view of "human nature" is false. Like nineteenth-century White European claims that colonialism is white man's burden or that racial hierarchies are part of the God-given natural order, a notion that capitalism is a natural expression of human nature is historically located and quite recent too. It represents the central tenet and excuse of today's reigning market ideology. This foundational untruth—what Klein calls "scapegoating 'humanity' for the greed and corruption of a tiny elite"[47]—is a rhetorical foundation of the ecocidal status quo and an entire critical tradition in climate scholarship has arisen to challenge it.[48] The real culprit behind climate change, this research shows, is not humanity but neoliberal capitalism, not the Anthropocene but "the Capitalocene."[49] Indeed, to displace the responsibility for ecocide from capitalism to human nature renders it impossible or irrelevant to engage in critical reflection about neoliberalism as the systemic driver of climate change. Thus, no matter what my students believe about human nature, I consider it critical for them to understand the nature of the socioeconomic system in which they live and how that system is the primary driver of ecocide.

IF IT'S THE SYSTEM, IS ANYONE RESPONSIBLE AT ALL?

Beyond the two activities mentioned earlier, I have used four other activities to facilitate discussions about the ecocidal nature of today's capitalism and the systemic causes of climate change. The first three are examples of using a children's book to help students grasp a specific concept. The fourth one is an experimental reader response activity I designed to tease out what a book is leaving unsaid and why it might matter.

Producerism

Critiques of capitalism and its excesses have long been centered on consumerism. While consumerism and especially the overconsumption of the wealthy few remain important issues to address, the focus on consumerism has paradoxically worked as another veiled excuse for the capitalist assault on the natural world—making it appear as if capitalism is a system that merely satisfies demand by providing people what they want. As many scholars have pointed out, however, capitalism is no longer a consumerist structure. It has evolved into a "producerist" system.[50] Its powerful, millions-of-dollars-worth advertising campaigns make us want things we don't really need—an experience most students are able to relate to. The indoctrination into teaching us to want more is not a response to our needs but a core operation of a system that needs to sell. The concept of consumerism, argues British ecophilosopher Rupert Read, "is a piece of false consciousness."[51] It misplaces responsibility from the system onto individuals, whose alleged "need" for products drives the system. The reality obscured by the rhetoric of mainstream economics, however, is that "Capitalism is a producerist system [whose] most brilliant product, . . . greatest achievement, [and] founding lie, is to produce individuals . . . ignorant of its real nature."[52] The key strategy of masking its nature as a producerist system is to make us believe that we are consumers and we live in a consumerist society. "So long as we think about ourselves as 'consumers,'" Read concludes, "we are blaming the victim."[53]

One excellent book that illustrates the producerist nature of capitalism is Dr. Seuss's *The Lorax* (1971). Although the story is most commonly interpreted as urging young people to take restorative environmental action in the wake of (adult/industrial) destruction, the overlooked gem of this story is how *The Lorax* represents the producerist nature of capitalism. When the Once-Ler arrives in the valley, they chop down the Truffula tree to produce a thneed, a ridiculously useless object that is *not* anything the Once-Ler claims it to be even though—potentially—one could wear it as a sock, a hat, a bicycle seat cover and so on. This potential usefulness of a thneed is then established as factual and blown beyond any proportions by the Once-Ler's credo: "A Thneed's a Fine-Something-That-All-People-Need!"[54] Ask your students to find textual evidence for how many customers actually purchase thneeds and whether we hear anything about why one might need a thneed in the first place. Then look for textual evidence for how the need for thneeds is manufactured through advertising and the Once-Ler's own claims—say, about why the factory is established and why the production must grow. Whose needs are being met? Do we ever see happy thneed customers? If not, who benefits from the production of thneeds? And who pays the environmental cost? This can be an eye-opening conversation, especially if you ask your students to collect evidence for the Once-Ler's shallow-sounding environmental concern—his claims "I meant no harm"[55] or "I've worried about it with all my heart"[56]—and contrast it with his actual actions, past and present. These conversations, incidentally, are places to introduce the concept of "greenwashing": corporate lip service to environmental care belied by their actions within neoliberal economy. Was the Once-Ler really unaware of the destruction? Or did they choose to ignore it? How many times did the Lorax petition the Once-Ler to stop the destruction, speaking on behalf of its victims: the trees, Bar-ba-Loots, Swomee-swans, Humming-fish? And how many times did the Once-Ler insist that the thneeds are absolutely necessary, so the needs of the victims are small in comparison? In this light, the supposedly repentant Once-Ler—who tells the story by first demanding payment(!)—is as hypocritical and destructive as the producerist capitalism they stand for. When they throw the last Truffula seed to the child character, the Once-Ler's urging to plant the trees—because now "Truffula Trees are what everyone needs"[57]—is less a call for environmental action than the dumping of the Once-Ler's devastation on the next generation. This is how producerism works: a few reap massive financial benefits for a while, while the many are left destitute in a desolate landscape of destruction that will continue for generations.

Extractivist Economy

The Lorax can also be used to illustrate another operational principle of neoliberal capitalism: its extractivist nature. In today's market economy, writes Vandana Shiva, "extractivism [is cloaked] as scientific, economic and human 'progress,'" which notion is then taken as "the basis of the domination, exploitation and colonization, enslavement and extraction."[58] As Macarena Gómez-Barris has shown in *The Extractive Zone* (2017), the violence of "extractivismo" is predominantly deployed against poor and Indigenous populations in the Global South in operations that "reduce, constrain, and convert life into commodities."[59] Shiva and Gómez-Barris are among many other scholars who have demonstrated that the logic of global capitalism is akin to that of a colonial invasion, in which a powerful group of outsiders descend on the local population to extract resources—think of James Cameron's *Avatar* (2009). The gains are then moved elsewhere while the devastation is left for the locals to deal with. How many executives of the several corporations that share ownership of the Canadian Tar Sands operation—the Syncrude Project, called "the world's most destructive oil operation"[60]—actually live within a few hundred miles of the apocalyptic devastation zone the project creates? How many appreciate the scale and scope of the damage? Or hear the voices of those marginalized, uprooted, and dispossessed by the destruction? For extractivismo, pollution and destruction are always someone else's problem. Extractivist economy operates on the assumption of "Cheap Nature," a notion that humanity is radically separate from nature and entitled to extract anything from it for its own benefit—in fact, the benefit of the few who control "the cash nexus."[61] And it requires a concept of "elsewhere," where the benefits of environmental demolition can be translated into money and resources to enjoy. This kind of "Windigo" or demonic economy, as Native American scholar Robin Wall Kimmerer calls it,[62] has been the story of capitalism so far—from rich white ladies enjoying leisurely lives in England paid for by "invisible" slave labor in the overseas plantations, to today's corporate moguls enjoying incomes, bonuses, and benefits hundreds of times higher than the underpaid, underinsured, and expendable workers in their corporations. However, the closer we get to planetary boundaries today, the more elusive this climate-safe elsewhere becomes. Recently, the elsewhere has been increasingly located in space, as Mars for Elon Musk. Or, in the flipside of the extractivist logic, billionaires such as Jeff Bezos have been suggesting offloading destruction to other planets while preserving what remains of the earth as a pleasure garden of sorts. "We have a choice," Bezos lures the audience, "we can have a life of stasis, because Earth is finite. . . . Or . . . , if we continue to grow . . . all the polluting industry will be done in space where we have infinite resources."[63] Compare this with the Once-Ler's: "I meant no harm. I most truly did not. // But I had to grow bigger. So bigger I got."[64] Equating respect for limits with stasis and projecting limitless growth with infinite resources as humanity's destiny are foundational claims of extractivist economy masquerading as "progress." It is the same extractivist logic, as Eileen Crist has argued, that is destroying Earth in the name of "delivering modern freedoms"—a suicidal project that prioritizes infinite growth "at the expense of the natural world" and "built out of its demolition."[65] In the words of Naomi Klein, "what the climate needs to avoid collapse is a contraction in humanity's use of resources; what our economic model demands to avoid collapse is unfettered expansion. Only one of these sets of rules can be changed, and it's not the laws of nature."[66]

Bill Peet's *The Wump World* (1970)—potentially paired with Peet's *Farewell to Shady Glade* (1966)—offers stories that illustrate the workings of extractive economy. In *The Wump World*, the green and clean world of the Wumps is invaded by the Pollutians from the Planet Pollutus. Represented as humanoid settlers, the Pollutians plant their flag and declare the planet theirs. "Soon the entire Wump World was overrun"[67] as giant machines begin transforming the landscape into a concrete jungle. The Wumps hide underground and their world becomes "one great turmoil of noise and confusion."[68] I have asked students to identify reasons why the Pollutians "improve their new wonderful world"[69] in such destructive ways. What drives them? What benefits do they get from the growth and development they create? How is their "economy" different from the Wumps' one? This can lead to eye-opening conversations about natural limits and sustainability, about the meaning of progress, about what it means to be uprooted, alienated, and dispossessed. In discussions, students noted that the only reason behind what the Pollutians do seems to be an urge for "more of everything."[70] And that the author "saves" the wumps by allowing them to hide and wait out the invasion. At some point, however, the planet's resources are consumed. Its water is undrinkable, its air is unbreathable. The Pollutians grow restless, petition their leader to find another planet, and when it is discovered they blast off. I ask students to describe the Wumps' feelings as they emerge from the underground caverns. I ask them to find phrases that describe the "improved" world left by the Pollutians and the brutalization of the excluded Wumps. Shiva notes that the ultimate result of extractivism is always "an ecological crisis"[71] and I invite students to consider how this end point is represented. The real question this book raises is what one can do in the wake of planetary destruction. *The Wump World* does not offer answers, but it helps us ask if we are behaving like space aliens. The book enables a reflection on the dark side of neocolonialist calls, like those of Bezos, for us to become "a dynamic entrepreneurial civilization in space."[72] Reading about the Pollutians helps students see that we have created a mechanized, space alien civilization that is ravaging our home planet as if we had many others available to us. The thing is, we don't. Unlike in *Farewell to Shady Glade*—in which a group of animals is displaced from their glade travel across the country to find a new home—the Wumps survive by hiding. But hiding or escape to another place are not the options we have. The only alternative to planetary devastation we have is to change our socioeconomic system.

Fungibility

Another foundational assumption of neoliberal capitalism is that everything is fungible. That is, everything can be monetized and translated into a financial value; that any problem we create can be paid its way out of. This assumption is almost never expressed directly but it lies at the heart of "an economic model based on the illusion of limitless growth,"[73] which commodifies everything in the world into tradeable goods that can be bought and sold, including future outcomes and value that are central to the operations of speculative finance. The assumption about fungibility is part of market fundamentalism that has "systematically sabotaged our collective response on climate change."[74] As long as it is cheaper to pollute in the present, there is no financial incentive to invest in services and infrastructure whose monetary returns are in the future or unclear. For example, how can one profit financially from clean air or "saving" Amazon forests? The way current market economy is organized makes it much easier, as Kim Stanley Robinson explains, to kick the costs of ecological restoration to the future generations—generations which, in the dispensational imagination of straight-line economics, will necessarily be much richer than we are. In this mainstream economic vision,

> if we were to hand down , three generations down the line, a devastated Earth, they would just pay the necessary money to recreate the Earth in a proper way. Unfortunately, everything is not fungible and you can't pay your way out of ocean acidification and a mass extinction event. Thus, no matter how rich future generations may be in economic terms, ecological collapse is not something that can be undone through payment.[75]

This notion that environmental destruction is not fungible—that is, that no amount of money can restore destruction that has crossed certain thresholds—is an important element of climate literacy. It adds radical urgency to the present efforts at conservation and helps reign in the "growthism" paradigm that drives today's economy like it does the Pollutians' and Once-Ler's enterprises. To help students grasp this concept, I have used Shaun Tan's short story "Bears with Lawyers" from *Tales from the Inner City* (2018). The story imagines a situation in which bears hire legal representatives to sue humanity in a class action of epic proportions. As the court procedures begin, it turns out that "Human Law is not the only legal system on the planet"[76] and that Bear Law supersedes Human Law in the cosmic hierarchy. The massive evidence bears present is "humbling, beautiful, indisputable, and horrifying"; it reveals that Human Law, including finance, "meant very little to the rest of the world."[77] But human crimes did. "Reading any fraction of the material, a case against humankind gathered over some ten thousand years, was an exercise in abject despair." It included both crimes humans recognize as well as crimes humans never even heard of, such as "Spiritual Exclusion."[78] The worst of it is that humans know the bears are right. After months of mining technicalities and loopholes, humanity is defeated. Humans offer bears a settlement, "a figure that made our eyes water and our mouths dry, a figure too staggering to ever make public."[79] If you divide the story into two parts, this is the place to pause for a conversation. Why didn't humanity realize the destruction we enacted on the biosphere? What makes us believe that human systems, views, and beliefs are the most important in the world? When the payday comes, will we be able to negotiate the settlement with the planet? What will the bears do? How will humans make up for their species' crimes? What recompense is adequate? The rest of the story explores these questions. I found it productive to design conversations in groups where one student knows the rest of the story and others have to imagine it. The learning here is that the bill for environmental destruction will come in kind. It will not be possible to settle it in cash.

On Personal Responsibility and the World the Rich (Un)Made

In teaching students about the systemic causes of climate change, one danger is that the more we stress how neoliberalism works, the easier it is to assume that planetary destruction is nobody's fault. If ordinary people are victims and unwilling accomplices of the global producerist Capitalocene, the origins and the blame for climate change fall on the system rather than on any particular humans. This position has to be nuanced. While the system is indeed the primary driver of ecocide—primary in a sense that no amount of change in individual consumer choices will ever be enough unless we transform our degenerative, divisive, and growth-addicted economy into one that is "regenerative, distributive, and able to thrive beyond growth,"[80] also primary in a sense that a mere ninety-one companies, acting as legal persons, are responsible for 78% of all greenhouse gas emissions[81]—the system has been created by decisions of specific people, is run by specific people, and benefits specific people. For example, executive decisions in the nineteen largest Big Oil companies in the world—companies whose record of environmental destruction has been calculated to be in the range of 35% of all energy-related CO_2 emissions worldwide[82]—are made by about 500 people.[83] Legislative (political) decisions that enable ecocide and executive (corporate) decisions in industries that support the ecocidal status quo are likewise made by a very, very, very small group of highly privileged people, all of them from the upper top 1% of income earners. Given the devastating impact of these decisions on the planet, it is increasingly problematic to argue that these people are merely "doing their job." While the nuances are many, there is a growing consensus that key people responsible for ecocide can be named and persecuted even within the existing international law.[84] There is also a recognition that a wealthy minority—one to 10% of the global population—is actively profiting from factors that contribute to climate change.[85] This small group is also disproportionately responsible for lifestyle consumption emissions.[86] So much so that the UN *Emissions Gap Report 2020* explicitly urges the global rich to rapidly reduce their carbon footprint by as much as 97%.[87]

Introducing the 1% (or 10%) lens allows us to talk about responsibility for climate change that is both systemic and personal. These discussions are extremely difficult because some of us and some of our students are members of this 1% to 10% group—thus the "bad people" disproportionately responsible for destroying the planet. I have yet to find a children's book that addresses affluence as a problem at the heart of climate change, but I have experimented with Geoff Rodkey's *We're Not from Here* (2019) to invite students into a discussion which helps them grasp the consequences of social inequality for our planet's survival. The premise of the novel is that Earth's biosphere collapsed and only 2,400 people made it out into space. While the collapse is not what the book is about, I created a questionnaire which asks students to imagine who was saved and to speculate about how Earth was made uninhabitable. The three students I was able to survey so far all pointed to issues of climate change. "The only reason I thought of while reading the book," wrote Elise (2020), "was that climate change got so out of control due to corporations that the planet became uninhabitable." I then followed with a number of other questions that the book leaves unsaid. Why did only 2,400 people get out, while the population at the time of collapse was said to be nine billion? Were any people left behind—if so, how many are in your estimate? How do you think the selection of evacuees was made? How many evacuees were people of color, Indigenous, First Nations, or from cultures/nations other than Europe and North America? Were any animals saved too? Do you think you or your family would be saved and why? What I have seen, even in this preliminary first run, is that the more questions students consider, the more they realize the absolute criminality of a situation in which "just a few bad people" are given the power to decide the fate of billions. For example, whereas initially students are sympathetic or very sympathetic to the plight of the refugees, once they begin to suspect that the refugees are families of the super wealthy whose actions contributed to the destruction of Earth, they are unsympathetic (two votes) or indifferent (one vote) to their fate. All three students were offended at the idea that the refugees claim to represent the human race to an alien civilization. All three also chose the same answer, "They would be least impacted," to the question "How would the wealthy be impacted in case of major climate catastrophes in the future?" I am currently building a more fine-tuned reader-response questionnaire for a larger group of respondents, but even this first experiment has showed me the effectiveness of asking questions about wealth disparity as they translate into responsibility for climate change and the chance to survive if/when a climate catastrophe occurs.

CONCLUSION

When young people say "It wasn't us!" they are right. It was us, their parents and grandparents. Unlike our grandparents—who really did not know about human impacts on the biosphere—our climate illiterate, arrogant, and market-obsessed generation has made the world safe for multinational corporations but extremely unsafe for our own children and grandchildren. In the past four decades, we have allowed unprecedented destruction of the biosphere, unprecedented concentration of wealth and power in the hands of the few, and unprecedented suffering of billions, especially in the Global South, whose lives have been devastated by the slow violence of accretive ecocide. We have failed for so long and so profoundly that we deserve young people's anger. And they deserve our full, radical support. We need to stand up with Greta Thunberg and other youth activists, whose fight now is for everyone's future. When Thunberg says that "the answers we seek will not be found within the politics of today,"[88] one of our challenges is to reimagine education, centering climate literacy as the foundation of an emerging ecological civilization. We should use our voices to raise awareness, our classrooms to mobilize action, and our pedagogy to connect students to movements. We can still turn this around, no matter how nearly impossible it seems.

NOTES

1. Geoff Rodkey, *We're Not from Here* (New York: Crown Books for Young Readers, 2019), 81.
2. Rodkey, 81.
3. Specific numbers are as follows: Children's Literature Association Quarterly: 2 out of 1,976 articles published to date, The Lion and the Unicorn: 2 out of 863, Children's Literature: 1 out of 732, Children's Literature in Education: 2 out of 1,264, International Research in Children's Literature: 0 out of 185, Journal of Children's Literature: 3 out of 292, Jeunesse: 2 out of 188, Barnboken: 2 out of 186, and Research on Diversity in Youth Literature: 0 out of 40.
4. Alexandra Panos and James Damico, "Less than One Percent is not Enough: How Leading Literacy Organizations Engaged with Climate Change from 2008 to 2019." *Journal of Language and Literacy Education*, 17, no. 1 (Spring 2021).
5. IPCC Special Report, "Global Warming of 1.5ºC," *Intergovernmental Panel of Climate Change*, October 8, 2018, https://www.ipcc.ch/sr15/. Accessed December 12, 2020.
6. Clare Farrell, Alison Green, Sam Knights, and William Skeaping, ed., *This Is Not a Drill: The Extinction Rebellion Handbook* (London: Penguin Books, 2019), 1.
7. See, for example, Lisa Kretz, "Emotional Solidarity: Ecological Emotional Outlaws Mourning Environmental Loss and Empowering Positive Change," in *Mourning Nature: Hope at the Heart of Ecological Loss and Grief*, edited by Ashlee Cunsolo Willox and Karen Landman, 258–291 (Montréal: McGill-Queen's University Press, 2017); Robert J. Brulle and Kari Marie Norgaard, "Avoiding Cultural Trauma: Climate Change and Social Inertia," *Environmental Politics*, 28, no. 5 (2019): 886–908. doi: 10.1080/09644016.2018.1562138; Panu Pihkala, "Eco-Anxiety and Environmental Education." *Sustainability* 12 (2020): 10149, 1–38; doi: 10.3390/su122310149; and Cathryn van Kessel, "Teaching the Climate Crisis: Existential Considerations." *Journal of Curriculum Studies Research*, 2, no. 1 (June 2020): 129–145.
8. Imre Szeman and Dominic Boyer, ed., *Energy Humanities: An Anthology* (Baltimore, MD: John Hopkins UP, 2017), 5.
9. See, for example, Elizabeth Kolbert, "Why Facts Don't Change Our Minds." *The New Yorker*, February 27, 2017. https://www.newyorker.com/magazine/2017/02/27/why-facts-dont-change-our-minds?; and Matthew J. Hornsey, "Why Facts Are Not Enough: Understanding and Managing the Motivated Rejection of Science,"

Current Directions in Psychological Science (November 12, 2020): 1–9, https://journals.sagepub.com/doi/10.1177/0963721420969364.

10. See, for example, Naomi Oreskes and Eric Conway, *Merchants of Doubt* (New York: Bloomsbury, 2010); Naomi Oreskes, *Why Trust Science?* (Princeton, NJ: Princeton University Press, 2019); and Philippe D. Tortell, "Earth 2020: Science, Society, and Sustainability in the Anthropocene." *PNAS*, 117, no. 16 (April 20, 2020). www.pnas.org/cgi/doi/10.1073/pnas.2001919117.

11. See George Marshall, *Don't Even Think about It: Why Our Brains Are Wired to Ignore Climate Change* (New York: Bloomsbury, 2014).

12. Timothy Morton, *Hyperobjects: Philosophy and Ecology after the End of the World* (Minneapolis, MN: University of Minneapolis Press, 2013), 1.

13. "Greenhouse Gases Factsheet," University of Michigan Center for Sustainable Systems. 2020. http://css.umich.edu/factsheets/greenhouse-gases-factsheet.

14. BBC, "Australia's Fires 'Killed or Harmed Three Billion Animals,'" July 28, 2020. https://www.bbc.com/news/world-australia-53549936.

15. Marek Oziewicz, *Justice in Young Adult Speculative Fiction: A Cognitive Reading* (New York, Routledge, 2015).

16. Philippe D. Tortell, "Earth 2020: Science, Society, and Sustainability in the Anthropocene," *PNAS*, 117, no. 16 (April 20, 2020), www.pnas.org/cgi/doi/10.1073/pnas.2001919117, 8688.

17. Rafaella Baccolini and Tom Moylan, eds., *Dark Horizons: Science Fiction and the Dystopian Imagination* (New York: Routledge, 2003), 2–5.

18. See, specifically, Fredric Jameson, *The Political Unconscious: Narrative as a Socially Symbolic Act* (Ithaca, NY: Cornell University Press, 1981); Naomi Klein, *This Changes Everything: Capitalism vs. the Climate* (New York: Simon and Schuster, 2014); and Peter Joseph, *The New Human Values Movement: Reinventing the Economy to End Oppression* (Dallas, TX: BenBella Books, 2017).

19. Matthew Schneider-Mayerson, "The Influence of Climate Fiction: An Empirical Survey of Readers," *Environmental Humanities*, 10, no. 2 (November 2018), 490.

20. Marek Oziewicz, "Planetarianism NOW: On Anticipatory Imagination, Young People's Literature, and Hope for the Planet," in *Pedagogy in the Anthropocene: Re-wilding Education for a New Earth*, edited by Michael Paulsen, Shé Mackenzie Hawke and Jan Jagodzinski (London: Palgrave, forthcoming, 2021).

21. Al Gore, *An Inconvenient Truth*, Dir. Davis Guggenheim (Paramount Classics, 2006).

22. IPCC Special Report, "Global Warming of 1.5°C."

23. Anna Kamenetz, "Most Teachers Don't Teach Climate Change; 4 in 5 Parents Wish They Did," *MPR News* (April 22, 2019), https://www.npr.org/2019/04/22/714262267/most-teachers-dont-teach-climate-change-4-in-5-parents-wish-they-did. Accessed December 11, 2020.

24. Melanie Burney, "In New Jersey Schools, Climate Change Education Will Be Mandatory," *The Philadelphia Inquirer* (July 2, 2020), https://www.inquirer.com/education/climate-change-new-jersey-mandatory-curriculum-tammy-murphy-20200702.html. Accessed December 13, 2020.

25. John B. King et al., "Open Letter to President-Elect Joseph Biden Vice President-Elect Kamala Harris" (November 18, 2020), https://www.k12climate-action.org/img/Biden-Transition-ED-Climate-Formatted.pdf. Accessed December 13, 2020.

26. NOAA, "Climate Literacy: The Essential Principles of Climate Sciences" (March 2009), https://aamboceanservice.blob.core.windows.net/oceanservice-prod/education/literacy/climate_literacy.pdf. Accessed December 17, 2020.

27. See, Krista K. Hiser and Matthew K. Lynch, "Worry and Hope: What College Students Know, Think, Feel, and Do about Climate Change," *Journal of Community Engagement and Scholarship*, 13, no. 3 (2021).

28. See, for example, Naomi Klein, *This Changes Everything: Capitalism vs. the Climate* (New York: Simon and Schuster, 2014); Imre Szeman and Dominic Boyer, eds. *Energy Humanities: An Anthology* (Baltimore, MD: John Hopkins UP, 2017); and Naomi Klein, *The (Burning) Case for a Green New Deal* (New York: Simon and Schuster, 2019).

29. Amitav Ghosh, *The Great Derangement: Climate Change and the Unthinkable* (Chicago: The University of Chicago Press, 2016), 154.

30. Bill McKibben, "Trust is Hard to Find at the U.N. Summit in Glasgow," The New Yorker, Nov 9, 2021, https://www.newyorker.com/news/daily-comment/trust-is-hard-to-find-at-the-un-climate-summit-in-glasgow.

31. Timothy Morton, *Hyperobjects*, 54.

32. Morton, 83.

33. Morton, 32.

34. See, for example, Naomi Oreskes and Eric Conway, *Merchants of Doubt* (New York: Bloomsbury, 2010) or David Kaiser and Lee Wassermann, "The Rockefeller Family Fund vs. Exxon," *The New York Review of Books* (December 8, 2016), https://www.nybooks.com/articles/2016/12/08/the-rockefeller-family-fund-vs-exxon/. Accessed December 10, 2020.

35. See, for example, Naomi Klein, *This Changes Everything: Capitalism vs. the Climate* (New York: Simon and Schuster, 2014); Amitav Ghosh, *The Great Derangement: Climate Change and the Unthinkable* (Chicago: The University of Chicago Press, 2016); Peter Joseph, *The New Human Values Movement: Reinventing the Economy to End Oppression* (Dallas, TX: BenBella Books, 2017); Imre Szeman, and Dominic Boyer, eds. *Energy Humanities: An Anthology* (Baltimore, MD: John Hopkins UP, 2017); and Rupert Read and Samuel Alexander, *This Civilization Is Finished: Conversations on the End of Empire—and What Lies Beyond* (Melbourne: Simplicity Institute, 2019).

36. Max Haiven and Alex Khasnabish, *The Radical Imagination: Social Movement Research in the Age of Austerity* (Halifax and Winnipeg: Fernwood Publishing, 2014), 99.

37. Kim Stanley Robinson, "Rethinking Our Relationship to the Biosphere," Bioneers Annual Conference: Revolution from the Heart of Nature (2015), https://www.youtube.com/watch?v=489I0gZlepM. Accessed December 15, 2020.

38. Rob Nixon, *Slow Violence and the Environmentalism of the Poor* (Cambridge, MA: Harvard University Press, 2011), 2.

39. Nixon, 37.

40. Naomi Klein, *This Changes Everything: Capitalism vs. the Climate* (New York: Simon and Schuster, 2014), 19.

41. Paul Krugman, "Why We're in a New Gilded Age," *The New York Review of Books* (May 8, 2014), https://www.nybooks.com/articles/2014/05/08/thomas-piketty-new-gilded-age/. Accessed December 13, 2020.

42. ASCE 2017 Infrastructure Report Card. American Society of Civil Engineers. 2021, https://www.infrastructurereportcard.org/. Accessed December 17, 2020.

43. Abigail Hess, "How Student Debt Became a $1.6 Trillion Crisis," *CNBC* (June 12, 2020), https://www.cnbc.com/2020/06/12/how-student-debt-became-a-1point6-trillion-crisis.html. Accessed December 17, 2020.

44. Dylan Matthews, "Here's How Much You Had to Make in 2017 to Pay More Income Tax Than Donald Trump," *Vox* (September 28, 2020), https://www.vox.com/policy-and-politics/21459782/trump-750-tax-returns-nyt. Accessed December 17, 2020; and David Leonhardt, "18 Revelations From a Trove of Trump Tax Records," *The New York Times* (November 13, 2020), https://www.nytimes.com/2020/09/27/us/trump-taxes-takeaways.html. Accessed December 17, 2020.

45. Geoff Rodkey, *We're Not from Here*, 81.

46. Rebecca Solnit, *Hope in the Dark: Untold Histories, Wild Possibilities*, 3rd ed. (Chicago, IL: Haymarket Books, 2016), xi.

47. Naomi Klein, *The (Burning) Case for a Green New Deal* (New York: Simon and Schuster, 2019), 252.

48. See, for example, Eileen Crist, *Abundant Earth: Toward an Ecological Civilization* (Chicago and London: The University of Chicago Press, 2019); and Jason Moore, ed., *Anthropocene or Capitalocene: Nature, History, and the Crisis of Capitalism* (Oakland, CA: PM Press, 2016).

49. Jason Moore, "The Rise of Cheap Nature." In *Anthropocene or Capitalocene: Nature, History, and the Crisis of Capitalism*, edited by Jason W. Moore (Oakland, CA: PM Press, 2016), 93.

50. Rupert Read and Samuel Alexander, *This Civilization Is Finished: Conversations on the End of Empire—and What Lies Beyond* (Melbourne: Simplicity Institute, 2019), 66.

51. Rupert Read and Samuel Alexander, 66.

52. Rupert Read and Samuel Alexander, 66–67.

53. Rupert Read and Samuel Alexander, 67.

54. Dr. Seuss, *The Lorax* (New York: Random House, 1971), Apple books, no page nos.

55. Dr. Seuss.

56. Dr. Seuss.

57. Dr. Seuss.

58. Vandana Shiva (with Kartikey Shiva), *Oneness vs. the 1%: Shattering Illusions, Seeding Freedom* (London: Chelsea Green Publishing, 2020), 18.

59. Macarena Gómez-Barris, *The Extractive Zone: Social Ecologies and Decolonial Perspectives* (Durham, NC: Duke University Press Books, 2017), xix.

60. Stephen Leahy, "This is the World's Most Destructive Oil Operation and It's Growing," *National Geographic* (April 11, 2019), https://www.nationalgeographic.com/environment/2019/04/alberta-canadas-tar-sands-is-growing-but-indigenous-people-fight-back/. Accessed December 19, 2020.
61. Jason Moore, "The Rise of Cheap Nature," 81.
62. Robin Wall Kimmerer, *Braiding Sweetgrass: Indigenous Wisdom, Scientific Knowledge, and the Teaching of Plants* (Minneapolis, MN: Milkweed Editions, 2013), 376.
63. Jeff Bezos, "Jeff Bezos Speaks at Amazon India Event," BloombergQuint (January 14, 2020), YouTube, https://www.youtube.com/watch?v=jzfXlg-wyUU, 13:32-14:25. Accessed December 21, 2020.
64. Dr. Seuss.
65. Eileen Crist, *Abundant Earth: Toward an Ecological Civilization* (Chicago and London: the University of Chicago Press, 2019), 5.
66. Naomi Klein, *This Changes Everything*, 21.
67. Bill Peet, *The Wump World* (Boston, MA: Houghton Mifflin, 1970), 17.
68. Peet, 21.
69. Peet, 18.
70. Peet, 24.
71. Vandana Shiva (with Kartikey Shiva), *Oneness vs. the 1%*, 18.
72. Jeff Bezos, "Jeff Bezos Speaks at Amazon India Event," 15:18.
73. Vandana Shiva (with Kartikey Shiva), *Oneness vs. the 1%*, 177.
74. Naomi Klein, *This Changes Everything*, 19.
75. Kim Stanley Robinson, "Climate Change Forces Post-Capitalism," Climate Change: Views from the Humanities. A Nearly Carbon Neutral Conference. UC Santa Barbara's Environmental Humanities Initiative, 2016, http://ehc.english.ucsb.edu/?page_id=13544, 9:00-9:47. Accessed December 14, 2020.
76. Shaun Tan, *Tales from the Inner City* (New York: Arthur A. Levine Books, 2018), 175.
77. Tan, 176.
78. Tan, 177.
79. Tan.
80. Kate Raworth, "A New Economics," in *This Is Not a Drill: The Extinction Rebellion Handbook*, edited by Clare Farrell, Alison Green, Sam Knights and William Skeaping (London: Penguin Books, 2019), 146.
81. Polly Higgins, "Ecocide Crime: A Crime of Our Time," *HagueTalks on Human Rights Day* (December 18, 2018), You Tube, https://www.youtube.com/watch?v=tgQ9kVzy1TM, 7:40-7:57. Accessed December 3, 2020.
82. Matthew Taylor and Jonathan Watts, "Revealed: the 20 Firms Behind a Third of All Carbon Emissions," *The Guardian* (October 9, 2019), https://www.theguardian.com/environment/2019/oct/09/revealed-20-firms-third-carbon-emissions. Accessed December 12, 2020.
83. Kim Stanley Robinson, *The Ministry for the Future: A Novel* (New York: Hachette Book Group, 2020), 30.

84. John Horgan, “Exposing the World’s Biggest Carbon Emitters,” *Scientific American* (June 2, 2020) https://blogs.scientificamerican.com/cross-check/exposing-the-worlds-biggest-carbon-emitters/, Accessed December 11, 2020, and Polly Higgins, “Ecocide Crime: A Crime of Our Time,” *HagueTalks on Human Rights Day* (December 18, 2018), You Tube, https://www.youtube.com/watch?v=tgQ9kVzy1TM, Accessed December 22, 2020.

85. See Thomas Piketty, *Capital in the Twenty-First Century* (Cambridge, MA: Harvard University Press, 2014); Peter Joseph, *The New Human Values Movement: Reinventing the Economy to End Oppression* (Dallas, TX: BenBella Books, 2017); Vandana Shiva (with Kartikey Shiva), *Oneness vs. the 1%: Shattering Illusions, Seeding Freedom* (London: Chelsea Green Publishing, 2020); Tim Gore, “Confronting Carbon Inequality,” Oxfam International, September 21, 2020. https://www.oxfam.org/en/research/confronting-carbon-inequality; and Thomas Piketty, *Capital and Ideology* (Cambridge, MA: Harvard University Press, 2020).

86. Nicholas Beuret, “Emissions Inequality—A Gulf Between Global Rich and Poor.” *Social Europe* (April 10, 2019), https://www.socialeurope.eu/emissions-inequality.

87. United Nations Environmental Programme, “Emissions Gap Report 2020,” (December 9, 2020), https://www.unep.org/emissions-gap-report-2020, Accessed December 22, 2020.

88. Greta Thunberg, *No One Is Too Small to Make a Difference* (New York: Penguin Books, 2020), 64.

REFERENCES

ASCE 2017. *Infrastructure Report Card*. American Society of Civil Engineers. 2021. https://www.infrastructurereportcard.org/.

Baccolini, Rafaella, and Tom Moylan, ed. *Dark Horizons: Science Fiction and the Dystopian Imagination*. New York: Routledge, 2003.

Beuret, Nicholas, “Emissions Inequality—A Gulf Between Global Rich and Poor.” Social Europe, April 10, 2019. https://www.socialeurope.eu/emissions-inequality.

Bezos, Jeff. “Jeff Bezos Speaks at Amazon India Event,” BloombergQuint, January 14, 2020, YouTube. https://www.youtube.com/watch?v=jzfXlg-wyUU.

Burney, Melanie, “In New Jersey Schools, Climate Change Education Will Be Mandatory,” The Philadelphia Inquirer, July 2, 2020. https://www.inquirer.com/education/climate-change-new-jersey-mandatory-curriculum-tammy-murphy-20200702.html.

Crist, Eileen. *Abundant Earth: Toward an Ecological Civilization*. Chicago and London: The University of Chicago Press, 2019.

Dr. Seuss. *The Lorax*. New York: Random House, 1971. Apple books.

Farrell, Clare, Alison Green, Sam Knights and William Skeaping, eds. *This Is Not a Drill: The Extinction Rebellion Handbook*. London: Penguin Books 2019.

Gómez-Barris, Macarena. *The Extractive Zone: Social Ecologies and Decolonial Perspectives*. Durham, NC: Duke University Press Books, 2017.

Gore, Al. *An Inconvenient Truth*. Dir. Davis Guggenheim. Paramount Classics, 2006.

Ghosh, Amitav. *The Great Derangement: Climate Change and the Unthinkable*. Chicago: The University of Chicago Press, 2016.

Haiven, Max and Alex Khasnabish. *The Radical Imagination: Social Movement Research in the Age of Austerity*. Halifax, Winnipeg: Fernwood Publishing, 2014.

Hess, Abigail. "How Student Debt Became a $1.6 Trillion Crisis," CNBC, June 12, 2020. https://www.cnbc.com/2020/06/12/how-student-debt-became-a-1point6-trillion-crisis.html

Higgins, Polly. "Ecocide, the 5th Crime Against Peace." TEDxExeter. May 1, 2012. https://www.youtube.com/watch?v=8EuxYzQ65H4.

———. "Ecocide Crime: A Crime of Our Time." HagueTalks on Human Rights Day. December 18, 2018, YouTube. https://www.youtube.com/watch?v=tgQ9kVzy1TM

Horgan, John. "Exposing the World's Biggest Carbon Emitters," Scientific American, June 2, 2020. https://blogs.scientificamerican.com/cross-check/exposing-the-worlds-biggest-carbon-emitters/.

IPCC Special Report. "Global Warming of 1.5ºC," October 8, 2018. https://www.ipcc.ch/sr15/.

Kaiser, David and Lee Wassermann. "The Rockefeller Family Fund vs. Exxon." The New York Review of Books, December 8, 2016. https://www.nybooks.com/articles/2016/12/08/the-rockefeller-family-fund-vs-exxon/.

Kamenetz, Anna. "Most Teachers Don't Teach Climate Change; 4 In 5 Parents Wish They Did." MPR News, April 22, 2019. https://www.npr.org/2019/04/22/714262267/most-teachers-dont-teach-climate-change-4-in-5-parents-wish-they-did.

Kimmerer, Robin Wall. *Braiding Sweetgrass: Indigenous Wisdom, Scientific Knowledge, and the Teaching of Plants*. Minneapolis, MN: Milkweed Editions, 2013.

King, John B. et al. "Open Letter to President-Elect Joseph Biden Vice President-Elect Kamala Harris," November 18, 2020. https://www.k12climateaction.org/img/Biden-Transition-ED-Climate-Formatted.pdf.

Klein, Naomi. *This Changes Everything: Capitalism vs. the Climate*. New York: Simon and Schuster, 2014.

———. *The (Burning) Case for a Green New Deal*. New York: Simon and Schuster, 2019.

Krugman, Paul. "Why We're in a New Gilded Age." The New York Review of Books, May 8, 2014. https://www.nybooks.com/articles/2014/05/08/thomas-piketty-new-gilded-age/.

Leahy, Stephen. "This Is the World's Most Destructive Oil Operation and It's Growing." National Geographic, April 11, 2019. https://www.nationalgeographic.com/environment/2019/04/alberta-canadas-tar-sands-is-growing-but-indigenous-people-fight-back/.

Leonhardt, David. "18 Revelations from a Trove of Trump Tax Records." The New York Times, November 13, 2020. https://www.nytimes.com/2020/09/27/us/trump-taxes-takeaways.html.

Matthews, Dylan. "Here's How Much You Had to Make in 2017 to Pay More Income Tax Than Donald Trump." Vox, September 28, 2020. https://www.vox.com/policy-and-politics/21459782/trump-750-tax-returns-nyt.

Moore, Jason. "The Rise of Cheap Nature." In *Anthropocene or Capitalocene: Nature, History, and the Crisis of Capitalism*, edited by Jason W. Moore, 78–115. Oakland, CA: PM Press, 2016.

Morton, Timothy. *Hyperobjects: Philosophy and Ecology after the End of the World.* Minneapolis, MN: University of Minneapolis Press, 2013.

Nixon, Rob. *Slow Violence and the Environmentalism and of the Poor.* Cambridge, MA: Harvard University Press, 2011.

NOAA. "Climate Literacy: The Essential Principles of Climate Sciences." March 2009, https://aamboceanservice.blob.core.windows.net/oceanservice-prod/education/literacy/climate_literacy.pdf.

Oreskes, Naomi. *Why Trust Science?* Princeton, NJ: Princeton University Press, 2019.

Oreskes, Naomi and Eric Conway, *Merchants of Doubt.* New York: Bloomsbury, 2010.

Panos, Alexandra and James Damico. "Less than One Percent Is Not Enough: How Leading Literacy Organizations Engaged with Climate Change from 2008 to 2019." *Journal of Language and Literacy Education*, 17, no. 1 (Spring 2021), 1–21.

Peet, Bill. *The Wump World.* Boston, MA: Houghton Mifflin, 1970.

Piketty, Thomas. *Capital in the Twenty-First Century.* Cambridge, MA: Harvard University Press, 2014.

Raworth, Kate. "A New Economics." In *This Is Not a Drill: The Extinction Rebellion Handbook*, edited by Clare Farrell, Alison Green, Sam Knights and William Skeaping, 146–154. London: Penguin Books 2019.

Read, Rupert and Samuel Alexander. *This Civilization Is Finished: Conversations on the End of Empire—and What Lies Beyond.* Melbourne: Simplicity Institute, 2019.

———. "Climate Change Forces Post-Capitalism." Climate Change: Views from the Humanities. A Nearly Carbon Neutral Conference. UC Santa Barbara's Environmental Humanities Initiative, 2016. http://ehc.english.ucsb.edu/?page_id=13544.

———. "Rethinking Our Relationship to the Biosphere." Bioneers Annual Conference: Revolution from the Heart of Nature, 2015 https://www.youtube.com/watch?v=489I0gZlepM.

———. *The Ministry for the Future: A Novel.* New York: Hachette Book Group, 2020.

Rodkey, Geoff. *We're Not from Here.* New York: Crown Books for Young Readers, 2019.

Schneider-Mayerson, Matthew. "The Influence of Climate Fiction: An Empirical Survey of Readers." *Environmental Humanities* 10, no. 2 (November 2018), 473–500.

Shiva, Vandana (with Kartikey Shiva). *Oneness vs. the 1%: Shattering Illusions, Seeding Freedom.* London: Chelsea Green Publishing, 2020.

Solnit, Rebecca. *Hope in the Dark: Untold Histories, Wild Possibilities*. 3rd ed. Chicago, IL: Haymarket Books, 2016.

Tan, Shaun. *Tales from the Inner City*. New York: Arthur A. Levine Books, 2018.

Taylor, Matthew and Jonathan Watts. "Revealed: The 20 Firms Behind a Third of All Carbon Emissions." The Guardian, October 9, 2019. https://www.theguardian.com/environment/2019/oct/09/revealed-20-firms-third-carbon-emissions.

Thunberg, Greta. *No One Is Too Small to Make a Difference*. New York: Penguin Books, 2020.

Tortell, Philippe D. "Earth 2020: Science, Society, and Sustainability in the Anthropocene." *PNAS* 117, no. 16 (April 20, 2020). doi: 10.1073/pnas.2001919117.

United Nations Environmental Programme. "Emissions Gap Report 2020." December 9, 2020. https://www.unep.org/emissions-gap-report-2020.

Chapter 2

Amitav Ghosh and Arundhati Roy on Climate Change

A Pedagogical Approach to Awakening Student Engagement in Ecocriticism

Suhasini Vincent

Postcolonial ecocriticism examines the relationship between humans and nonhuman communities and strives to imagine new ways in which these ecologically connected entities can coexist, transform, and reconfigure themselves. Even though postcolonial environmental issues are often linked to the colonizer's invasion, settlement and depletion of natural resources, forced migration of peoples, fretting of animals and plants throughout the European empires, the ecosystem challenges that are faced by postcolonial countries today are due to non-egalitarian government policies and corporate-capitalist dominance. Students today are aware of the North-South divide and wonder if there are means of reconciling the Northern environmentalism of the rich and the Southern environmentalism of the poor. History textbooks reveal the ever-present ecological gap between the colonizer and colonized and the disparities that continue to exist.

Ecocritical readings as seen through the lens of Amitav Ghosh[1] and Arundhati Roy[2] reveal the need to bring postcolonial and ecological issues together and challenge imperialist modes of social environmental supremacy.

Both writers have considered the complex interplay of politics and corporate capitalism in the use of water, land, energy, and habitat. Their fictional and nonfictional ecocritical writing trace the social, historical, political, economic, and material coordinates of forests, rivers, bioregions, and species. It is interesting for a student to explore the myriad relationships between material practices and cross-cultural contexts in the postcolonial world. For example, an American student's curiosity would be awakened by how day-to-day life is lived in postcolonial India. This chapter will reveal how futures are governed in the developing world. An ecocritical reading of works by Ghosh and Roy will examine how the policies of decision-makers and ideas from proponents of climate change can lead to a transference from a "red" to a "green" politics. The field of ecocriticism is large as it encompasses domains of economics, anthropology, history, geography, geopolitics, and environmental science. Thus, this chapter will instigate students in high school and college levels, as well as researchers in different disciplines, to analyze texts by considering how the authors advocate different means of dwelling authentically and responsibly on our planet Earth.

Ghosh's *The Great Derangement: Climate Change and the Unthinkable* aims at exploring how long-term ecological crises have been ignored in serious fictional writing.[3] Ghosh wisely reflects on how "different modes of cultural activity: poetry, art, architecture, theater, prose fiction . . . have responded to war, ecological calamity, and crises of many sorts: why then, should climate change prove so peculiarly resistant to their practices?"[4] He muses further, "What is it about climate change that the mention of it should lead to the banishment from the preserves of serious fiction? And what does this tell us about culture writ large and its patterns of evasion?"[5] He foretells a scenario of a world where cities' forests like the Sundarbans would be replaced by seascapes, megalopolises would become uninhabitable, future generations would try to find answers in literature and art for the damaged world they had unwittingly inherited. He imagines future generations concluding that our present generation is rather deranged and oblivious to the hazards, risks, vulnerabilities, and perils of climate change:

> In a substantially altered world, . . . when readers and museum goers turn to art and literature of our time, will they not look, first and most urgently, for traces and portents of the altered world of their inheritance? And when they fail to find them, what should they—what can they—do other than to conclude that ours was a time when most forms of art and literature were drawn into the modes of concealment that prevented people from recognizing the realities of their plight . . . this era which so congratulates itself on its self-awareness, will come to be known as the time of the Great Derangement.[6]

In this work of nonfiction, Ghosh probes into possible reasons for the postcolonial writer's imaginative failure in the face of overwhelming evidence. He examines the inability of the present generation to grasp the violent scale of climate change and posits that this is reflected in the literature of our time, in the recording of history and in the political ambience of our day. Ghosh posits that the extreme nature of today's climate changes has resulted in making writers immune and resistant to contemporary modes of thinking and imagination. He points out that certain phenomena like destructive storms and meandering rivers do not figure in serious literary fiction but get relegated to other genres like science fiction or fantasy. Ghosh suggests that politics has suffered the same fate as literature and has become a matter of personal moral reckoning rather than an arena of collective action. He further argues that to limit fiction and politics to individual moral adventure comes at a great cost. Through his writing Ghosh has shown how fiction is the best of all cultural forms to voice ecocritical awareness.

Arundhati Roy's political nonfiction[7] has often been viewed with a disdainful eye[8] and considered as "objectionable writings."[9] In her essay,[10] Roy voices dissent against repression, globalization, economic progress, environmental exploitation, and dams proposed in the name of greater common good. Roy's political essays insist on the need to make a "material turn" and consider how materialistic activity in the name of progress affects human and nonhuman environments. The term "material turn" is used by Serenella Iovino in her work *Material Ecocriticism* in which she describes an enterprise of writing that is replete with "a material mesh of meanings, properties and processes, in which human and non-human players are interlocked in networks that produce undeniable signifying forces."[11] Roy's essays challenge government policies, discuss varying viewpoints of global and local concerns in India, criticize corporate philanthropy, and propose a new ecocritical perspective. Through her essays Roy expresses distrust of materialistic forces and encourages readers to take part in the new emerging paradigm of making a "material turn," thus considering possible ways of avoiding the depletion of our planet's resources through ecocritical advocacy and aesthetics. But one wonders why she writes about climate change in nonfictions, opting for the mode of the essay. Amitav Ghosh echoes the same idea by insisting that this practice is not due to a lack of information:

> A case in point is the work of Arundhati Roy: not only is she one of the finest prose stylists of our time, she is passionate and deeply informed about climate change. Yet all her writings on these subjects are in various forms of nonfiction.[12]

Ghosh also provides potential reasons for this sleight of writing by contending that "the discrepancy is not the result of personal predilections: it arises out of peculiar forms of resistance that climate change presents to what is now regarded as serious fiction."[13] In her "Arthur Miller Freedom to Write" lecture, Roy poses a series of questions on what it means to be a writer in today's world:

> So, as we lurch into the future, in this blitzkrieg of idiocy, Facebook "likes," fascist marches, fake-news coups, and what looks like a race toward

> extinction—what is literature's place? What counts as literature? Who decides? Obviously, there is no single, edifying answer to these questions.[14]

By musing on who decides, she concludes that there are no enlightening answers to these questions. She posits that the time is right to think together about a place for literature and the role it will play in climate change. In the same lecture, Arundhati Roy muses on what it means to be an "activist-writer" in a world where the delicate web of interdependence of Man and Nature is dictated by Capitalism and International treaties. Roy's essays in India are often regarded with a baleful eye as she disagrees with political decision-making, arouses dissent among the youth, and writes forcefully on topics other writers do not explore. She acknowledges that some did not count her nonfiction as writing:

> The writing sat at an angle to what was conventionally thought of as literature. Balefulness was an understandable reaction, particularly among the taxonomy-inclined—because they couldn't decide exactly what this was—pamphlet or polemic, academic or journalistic writing, travelogue, or just plain literary adventurism? To some, it simply did not count as writing. "Oh, why have you stopped writing? We're waiting for your next book."[15]

Though some considered her a mere "pen for hire,"[16] others were open to the call of change as her essays in the form of pamphlets were translated into other Indian languages, distributed freely in villages that were under attack, on university campuses where students were realizing that they were being lied to. The readership concerned those on the frontlines who understood her brand of literature.

Both Arundhati Roy and Amitav Ghosh believe that literature contributes to change and that it is built by writers and readers. Literature in their opinion occupies a fragile place that needs constant shelter and nurturing. With a pedagogical approach that harnesses the attention of readers, arouses ecocritical awareness of climate change, and prepares the youth of today to conserve our planet for the next generations, the present generation may not be doomed to the derangement Ghosh laments.

INTRODUCTION: EXPLORING THE WRITER'S LIVED EXPERIENCE

Postcolonial writers mine their own lived experiences to create a fictional ground that reflects the reality of their times. Amitav Ghosh has a doctorate in social anthropology[17] and his novels bear ample evidence of the ethnographic research and fieldwork he has conducted. His fictionalized narratives often draw from his own personal experience of migration and displacement. Arundhati Roy, an architect[18] involved in political activism, is a spokeswoman of the downtrodden and vehement critic of neo-imperialism,[19] fascism, corruption, and communalism. Her essays aim at laying bear the shortcomings in governmental decision-making and inspiring the youth of today to react. While using the works of Ghosh and Roy in a classroom, teachers should guide students to explore the fictional space by encouraging an analysis of select quotations referred to in this chapter. Teachers could choose a section for detailed study, surf the authors' websites[20] which contain excerpts of newspaper reviews and authors' notes. They could complete a detailed reading of a work and analyze it through the following three-step lens: figuration of the natural and material lay of the land; configuration of the writer's inherited experience in the form of stories, myths, legends, culture, and art; and the reconfiguration of the fictional or nonfictional space to inspire young readers to engage in the politics of climate change.

Mining and Recasting Childhood Stories for the Fictional Ground

In *The Great Derangement*, Amitav Ghosh mines his own experience, draws inspiration from stories of his childhood, and reconfigures the fictional ground. As he notes,

> No less than any other writer have I dug into my own past while writing fiction. . . . In essence, narrative proceeds by linking together moments and scenes that are in some way distinctive or different: these are, of course, nothing other than instances of exception. . . . It is through this mechanism that worlds are conjured up, through everyday details, which function "as the opposite of narrative." It is thus that the novel takes its modern form, through "the relocation of the unheard-of toward the background . . . while the everyday moves into the foreground."[21]

He recalls childhood family accounts of his ancestors who had been ecological refugees displaced from their roots on the banks of the River Padma[22] in the mid-1850s. When the mighty river on a whim decided to change course, the few vagrant inhabitants traversed forests and dry land moving westward to settle once again on the banks of another civilization-housing river, the Ganges.[23] Ghosh warns us that the river, a stable presence in the lives of his forefathers, had transformed into a meandering force that could not be relied on, nor taken for granted like the air we breathe. In his novel *The Hungry Tide*, Ghosh writes of the fickle nature of the landscape where mangrove forests of the Sundarbans[24] appear and disappear, merge and submerge, surprise and disrupt human lifestyles. He muses in *The Great Derangement*, "Even a child will begin a story about his grandmother with the words: 'in those days the river wasn't here, and the village was not where it is.'"[25] The fictional ground of *The Hungry Tide* is set in the fickle tidal landscape or *bhatir desh*[26] of the Sundarbans where the passage of the ebb tide leaves an ever-mutating and unpredictable terrain with "no borders to divide fresh water from salt, river from sea as the water tears away entire promontories and peninsulas; at other times it throws up new shelves and sandbars where there were none before."[27] In this indeterminate fluid fictional space of the Sundarbans, Ghosh takes the reader-voyager on a literary journey across the Gangetic delta. The voyage reveals the varied quests, travels, expeditions, and voyages of the protagonists in the eco-narrative. The mighty River Hooghly[28] dictates the literary journey in the novel as it meanders, changes course, reshapes land before sunrise, and reconfigures them with new paths before sunset. During its ever-shifting course, the River Hooghly intertwines with the River Meghna,[29] thus setting the theme of transformation and displacement for the human and animal inhabitants who adapt to the capricious river trajectories and deal with the challenging task of naming new, fresh, ever-emerging islands each day.

The character of Fokir, a local fisherman, recites a "legend passed on from mouth to mouth, and remembered only in memory."[30] The legend of Bon Bibi[31] features the relationship between human settlers and tiger predators in the Sundarbans. It describes an epoch when Bon Bibi had the divine task of rendering the Sundarbans fit for human inhabitation. In a battle between "good" and "evil," Bon Bibi emerges triumphant, divides the country of eighteen tides into two inhabitable zones for "humans" and for "demon-tiger" hordes of Dokhin Rai,[32] thus changing the time-set laws of the "survival of the fittest" to a new-named "law of the forest." The rich and greedy are aptly punished while the poor and righteous reap just rewards. In Fokir's river song, Dokhin Rai takes the form of the Ganges tiger with an insatiable craving for human flesh, and the fearsome tiger stalks Dhukey, a fisher boy, thus chanting an ever-living myth to affirm the hostilities between "death-bearing," "demon" tigers and "precious," "precarious" humans. Through the inclusion of the myth of Bon Bibi, Ghosh sets his novel in a twofold time zone of the present and the past that is influenced by a mythical time of incessant mutiny between "humans" and "animals," both species being constantly threatened by the vagaries of the river and the sea tide. By using the recurring chant of the Bon Bibi song, Ghosh highlights how myth influences and affects people's existence in the Sundarbans. Through the inclusion of the myth of Bon Bibi, he portrays the Sundarbans as a living entity endowed with the capacity to nurture "human" and "animal" protagonists who claim equal land, space, and territory thanks to an entitlement to land that can be traced back to roots in myth. Creation myths, Nature fables, and Family Stories exist in many cultures and families. Teachers can ask students to engage in an investigation or research project of such myths, legends, anecdotes, and stories that may have been passed down to them through their parents or grandparents. Like Ghosh who remembers his grandmother's accounts of meandering rivers and migration induced by a quest for other civilization-housing river zones, students may unearth similar tales of lost forests, farmlands that now accommodate skyscrapers, changing weather conditions, disappearing plant and animal life, and changing relationships with our planet Earth.

Figuring the Material Lay of the Land and Mining the Arts for Material Gain

In her second novel *The Ministry of Utmost Happiness*,[33] Roy writes, "People—communities, castes, races and even countries—carry their tragic histories and their misfortunes around like trophies, or like stock, to be bought and sold on the open market."[34] In her essay "Power Politics," Roy shows how the ruling elite in India, bureaucrats and multinational companies, welcome foreign dignitaries and present India as an investment conducive haven. Roy describes how Delhi was transformed, scrubbed, and cleaned to welcome Bill Clinton and his delegates in March 2000: "Whole cities were superficially spruced up. The poor were herded away, hidden from the presidential gaze. Streets were soaped and scrubbed and festooned with balloons and welcome banners."[35] In *Capitalism: A Ghost Story*, Roy criticizes the Indian government strategy of creating an aura of "Good Investment Climate"[36] and describes how Rand and Enron ended up owning Indian earth, air, and water. She questions the motives behind the drive to beautify Delhi for the Commonwealth Games when ephemeral laws were passed to present Delhi as a beautiful capital without any marring disfigurement. She describes how

> street vendors disappeared, rickshaw pullers lost their licenses, small shops and businesses were shut down. Beggars were rounded up, tried by mobile magistrates in mobile courts, and dropped outside the city limits. The slums that remained were screened off, with vinyl billboards that said DELHIciously Yours.[37]

Through metaphors of cleansing, Roy highlights how the capital city was purged of unwanted elements that marred the perfect scene staged to impress foreign investors. By eliminating the presence of street vendors, beggars, slums, and the poor, Delhi was reconfigured into a city that was inviting for foreign investment.

Further, Roy remarks on how main mining conglomerates mine the Arts—film, art installations, and literary festivals—to gain popularity as patrons of artistic and literary enterprise. She gives instances of how they exploit the arts for material gain: the bauxite mining conglomerate Vedanta sponsors a film competition named "Creating Happiness" encouraging the youth to make films on sustainable development;[38] Jindal Group,[39] the stainless steel giant, has a contemporary art magazine and supports artists who work with the same medium of stainless steel; Indian energy giant Essar, despite its mining scandal,[40] boasts of being the principal sponsor of the Tehelka Newsweek Think Fest[41] with its mind-stirring discussions by major writers and activists; the involvement of Tata Steel[42] and Rio Tinto[43] in the Jaipur Literary Festival[44] where the right to free speech in the style of Salman Rushdie is much publicized as voices "talking about the house, about everyone in it, about everything that is happening and has happened and should happen."[45,46] Despite the public display of embracing the arts, these miners of resources both natural and creative ignore "journalists, academics and filmmakers working on subjects unpopular with the Indian Government."[47]

Exploiting India for Material Gain

Postcolonial writing aims at revealing areas of colonization and its impact on the contemporary postcolonial world. The fiction and nonfiction of Amitav Ghosh and Arundhati Roy reveal the ambivalent themes and issues of colonial legacy that have shaped the so-called developing BRIC[48] countries of today. Ghosh's fiction examines the material gain of the colonists, and analyzes the dynamics of colonial trade, including its environmental impact on the indigenous population of colonial times. Arundhati Roy's fiction bears on the contrary less reference to colonial times, but gives a stark account of contemporary power politics, where richer countries dominate in international forums as experts on environmental development. She highlights also the imbalance of power politics in India where privatization has resulted in material wealth being held in the hands of a few megacorporations that thrive on material gain. While presenting the work of postcolonial authors, it is thus imperative to have a knowledge of the colonial past and engage in an exploration of the Opium Triangular Trade conducted by the British East Indian Company, which can be an eye-opener for students. The contemporary power politics in the Indian postcolonial context may be investigated by researching each one of the megacorporations mentioned in the ensuing sections.

The Opium Triangular Trade of the British East India Company

In the Ibis Trilogy, Ghosh gives an account of the materialistic intent behind the British East India Company's opium trade in India and their trade settlement in the Chinese province of Canton in the nineteenth century. Ghosh's eco-narrative depicts the resistance mounted by the Chinese authorities in protest against the lucrative British trade in the Golden Triangle.[49] The Golden Triangle refers "to a broadly triangular area with vertices with Burma, Laos and Thailand, where opium production was concentrated."[50] Ghosh's fictions highlight the empire-sized fortunes earned by the materialistically oriented British traders and pictures the suffering faced by the indigenous Indian and Chinese populations who stagger under the stupor of the potent somniferous drug of opium.

In the first novel of the Ibis Trilogy, *The Sea of Poppies*, the Black Water of the Sundarbans in the novel is seen as an expanse of water without "a boundary, a rim, a shore, to give it shape, and hold it in place."[51] Not only is the countryside blanketed with the parched remains of the poppy harvest, the debris of poppy leaves, dumped on the shore, makes its way to the Black Andaman Sea.[52] Through the drastic change in the landscape from rich and fertile irrigated plains to that of swamps and marshes with no potable water, Ghosh highlights the effects of greedy colonial enterprise on nature. In the second novel, *River of Smoke*, Ghosh's narrative takes the action forward from the cities, harbors, and plains of India to the Chinese trading outpost and opium destination of Canton. In the tiny foreign enclave of Fanqui-town,[53] the Cantonese outpost is populated by traders of the British East India Company[54] and surrounded by a flotilla of boats that ferry smuggled goods and serve as eating and pleasure houses. The West's fascination for the East as seen in the Orientalist-initial quest for spices like nutmeg, cloves, pepper, coffee, cacao, sugar, and tea lead eventually to the discovery of the addictive narcotic drug "opium." In *Opium, Empire and the Global Political Economy*, Carl Trocki posits that

> the British Empire, the opium trade, and the rise of global capitalism all occurred together. . . . Gold and silver from the West crossed the Atlantic or the Pacific, it ultimately found its way to Asia (east of Suez) to purchase the "riches of the East" and to allow the otherwise deprived inhabitants of the northwest Eurasian peninsula to share in the fabled Oriental splendours.[55]

This "ineradicable distinction between Western superiority and Oriental inferiority"[56] described by Edward Said[57] is evident in Ghosh's postcolonial textual response in the Ibis Trilogy where the margins write back and lay bare materialistic colonial intent of the past by reconfiguring and reliving the opium trade in the fictional space.[58] In *Flood of Fire*, the final book in the Ibis Trilogy, Ghosh describes how the British mobilized force on a large scale, unleashing the firepower of their advanced warship called Nemesis:[59]

> Baboo Nob Kissin raised a hand to point to the Nemesis, which was steaming past the burning forts, wreathed in dark fumes. Dekho—look: inside that vessel

> burns the fire that will awaken the demons of greed that are hidden in all human beings. That is why the British have come to China and Hindustan: these two lands are so populous that if the greed is aroused, they can consume the whole world. Today that great devouring has begun. It will end only when all of humanity, joined together in a great frenzy of greed, has eaten up the earth, the air, the sky.[60]

The novel is aptly named a *Flood of Fire*, as fire engulfs the Indian vessels. The incessant flood of fire prompts the Indian soldier Kesri to reflect, "So much death; so much destruction—and that too visited upon a people who had neither attacked or harmed the men who were so intent on engulfing them in this flood of fire."[61]

Ghosh's fictional enterprise of reliving the colonial trade is typical of what Ashcroft, Griffiths, and Tiffin claim to be "the rereading and the rewriting of the European historical and fictional record" (that) "is a vital and inescapable task at the heart of the postcolonial enterprise."[62],[63] He depicts how the British citizens of a global imperium dominate the world through a false doctrine of free trade. Ghosh shows that "despite all their cacklings about Free Trade, the truth was that their commercial advantages had nothing to do with markets or trade or more advanced business practices—it lay in the brute firepower of the British Empire's guns and boats."[64] During this era of triangular trade that enabled the three-legged journey of exchanging slaves in Africa for guns and brandy, the Middle Passage across the Atlantic to sell the slaves in the West Indies and North America, and the final cargo transportation of rum and sugar to England, opium served as the means of conducting business. It should be remembered that colonial trade conducted by these British opium barons contributed to colonial empire-building, large-scale opium production, and trading in the colonies, which led to the creation of the Golden Triangle[65] of Burma, Laos, and Thailand and the Golden Crescent of Afghanistan, Iran, and Pakistan.[66]

Privatization and the Dynamics of Power Politics

In *Capitalism: A Ghost Story*, Roy reminds readers that 800 million impoverished Indians "live side by side with spirits of the nether world, the poltergeists of dead rivers, dry wells, bald mountains and denuded forests."[67] This enumeration of dead resources essential for the survival of mankind serves to highlight significant figures: while 800 million impoverished Indians live on an average of 20 rupees per day, the top 100 richest industrialists in India possess wealth that amounts to one-fourth of the country's GDP of $300 billion. Material wealth in India is concentrated in the hands of a few. Roy juxtaposes her enumeration of depleted resources with the assets of India's richest billionaire Mukesh Ambani, whose twenty-seven-storey residence houses "three helipads, nine lifts, hanging gardens, ballrooms, weather rooms, gymnasiums, six floors of parking, and the six hundred servants."[68] Ambani's Reliance Industries Limited is typical of the market scenario in India where corporations like Tatas, Jindals, Vedanta, Mittals, Infosys, Essar are engaged in a race for supremacy in India's private sector. As Roy notes,

> They own mines, gas fields, steel plants, telephone, cable TV and broadband networks, and run whole townships. They manufacture cars and trucks, own the Taj Hotel chain, Jaguar, Land Rover, Daewoo, Tetley Tea, a publishing company, a chain of bookstores, a major brand of iodized salt and the cosmetics giant Lakme. Their advertising tagline could easily be: You Can't Live without Us.[69]

Roy labels this corporate race as "Gush-Up Gospel" and posits that "the more you have, the more you can have."[70] Roy details that the motive behind the foreign visits was to sweet-talk India into importing products which the country could manufacture on its own. In her essays, Roy warns readers that foreign investors were stalking big, varied game in the form of dams, mines, telecommunication, public water supply, and the dissemination of knowledge.

Roy notes that in the present trend of privatization, "India's new mega-corporations Tatas, Jindals, Essar, Reliance, Sterlite, are those who have managed to muscle their way to the head of the spigot that is spewing money extracted from deep inside the earth."[71] She contends that massive corporations, both multinational and domestic, are the principal agents of the concentration of wealth, and exploit the government's present policy of creating Special Economic Zones to have access to precious land and resources, resulting in the dispossession of millions. Whole scale privatization has spurred massive corruption and resulted in massive displacement of poor people who were promised inexistent jobs. She explains that with reference to Water Politics where "all over the world, weak, corrupt, local governments have helped Wall Street brokers, agro-business corporations and Chinese billionaires to amass huge tracts of land."[72] There is also another reference in Roy's essay "Power Politics," where she comments on the farce behind the meeting of International World Water Forum in March 2000 at The Hague. While 3,500 proponents like bankers, businessmen, politicians, economists, and planners pressed for the privatization of water, a handful of activist-opponents expressed dissent against the appropriation of a national resource by private multinationals. While the forum engaged in upholding false interest in issues regarding "women's empowerment, people's participation, and deepening democracy,"[73] the need to preserve the life-giving resource was pushed to the background. She exposes the existence of a lobbying effort at

The Hague, where cliques of consultants specialized in preparing dossiers for the Third World presented concocted data and virtual facts and figures. She states that they "breed and prosper in a space that lies between what they say and what they sell."[74] She thus shows how privatization thrives through the incomprehensible jargon of government dossiers and hostile takeover bids. Roy explains that privatization of India's mountains, rivers, and forests necessarily entails war, displacement, and ecological devastation. She refers to the signing of MOUs by the state governments of Chhattisgarh, Orissa, and Jharkhand[75] in 2005 when private corporations mined material gain worth trillions of dollars from mining bauxite, iron ore, and other minerals for "a pittance, defying even the warped logic of the Free Market. (Royalties to the government ranged between 0.5% and 7%)."[76] Her essay is written with the intention of protecting India's power reserves and seeks to reveal the dynamics of power politics to the world at large.

ENCOURAGING SOCIAL AND ENVIRONMENTAL ADVOCACY IN LITERATURE

Postcolonial ecocritical writing focuses on the political and geopolitical tensions that exist between the colonizer and the colonized during the twentieth century. Amitav Ghosh's fictional account reveals the efforts of colonists to create material prosperity detrimental to the colonized; on the other side, he configures how eco-friendly communities sprout in the fictional space to discover, identify, protect, and conserve threatened interconnected ecosystems and eco-societies. While exploring Ghosh's environmental advocacy, students should be encouraged to analyze how his fictional enterprise falls within the sphere of postcolonial ecocriticism that considers the problems of conserving biodiversity and distrusts the grandeur of empty materialistic quests. An ardent activist, Arundhati Roy's writing is a result of her fieldwork and involvement in environmental issues. She criticizes the Indian government's strategy of engaging in development at the cost of uneducated villagers' losing their livelihood and land. While discovering Roy's account of postcolonial India, teachers should encourage students to analyze how literature, both fictional and nonfictional, hold the key to environmental advocacy as it reveals the author's perspective of how development is not always for the greater common good. Teachers may encourage students to investigate their own country's governmental practices, policies, and decision-making. There may be recorded instances of how a government's environmental strategies created more harm than good. The goal is not to resurrect the forgotten stories of history to create shame or throw blame on students for inaction but to encourage a more contextualized understanding of injustice. This pedagogical approach aims at discovering the alignment of a green paradigm through an eco-narrative and environmental critical advocacy. This attempt to unite fictional aesthetics and advocacy is typical of postcolonial ecocriticism that seeks to advocate social and environmental justice in today's postcolonial world.

Establishing a Green Paradigm through an Eco-narrative

In Ghosh's eco-narrative *River of Smoke*, the troubled waters of the South China Sea witness the tryst between two storm-tossed vessels: the Anahita, a sumptuously built cargo ship laden with opium, and the Redruth, a two-mast vessel with a Cornish botanist and his assistant Paulette, an orphan who collects rare flora during the stormy journey. Paulette catalogues the Plants of Bengal and contributes to the body of collected knowledge called the *Materia Medica.*[77] She joins forces with the protagonist, a famous plant hunter, Fitcher Penrose to search for a rare camellia[78] and seek newer vistas through botanical exploration. Ghosh also pens the route of pilgrimage undertaken by early horticulturists to the Pamplemousses garden in his eco-narrative. He writes of the existence of a chaotic botanical garden where a wild and tangled muddle of greenery showed the existence of a primeval jungle "where African creepers were at war with Chinese trees, nor one where Indian shrubs and Brazilian vines were locked in a mortal embrace. This was a work of Man, a botanical Babel."[79] Thus, Ghosh reconfigures new spaces of postcolonial identity through an eco-narrative that enables the reader to identify himself with the central premise of articulating resistance against materialistic forces. Paulette and Penrose voice the need to protect greenery and nature from forces that strive to destroy the landscape. As a child of nature, Paulette had been taught by her father Pierre Lambert to love nature and consider it as a kind of spiritual striving whereby the quest was to comprehend the inner vitality of each species:

> If botany was the Scripture of this religion, then horticulture was its form of worship: tending a garden was, for Pierre Lambert, no mere matter of planting seeds and pruning branches—it was a spiritual discipline, a means of communicating with forms of life that were necessarily mute and could be understood only through a careful study of their own modes of expression—the languages of efflorescence, growth and decay: only thus he had taught Paulette, could human beings apprehend the vital energies that constitute the Spirit of the Earth.[80]

During the passage through the North China Sea, Paulette identifies a large variety of plants and tends them like a priestess performing a spiritual ritual.

Ghosh lays stress on the interconnected nature of different life-forms and Paulette's quest to name unnamed flora from Chinese territory that can be considered as a creative endeavor to consider how these ecologically connected groups can be creatively transformed. The mundane tasks of planting, pruning, and watering them become acts of discipline and a means of communication with mute forms of existence that manifest different kinds of vital energies that constitute the spirit of the earth. She carefully observes the procedures and protocols on board during times of storms. At the same time the eco-conservationists in the fictional space have to deal with the discontent of the seamen on board who regard the plants as threats to their existence and deny them water in times of scarcity or empty the pots of precious water when menaced by storms. The reader encounters a multiplicity of voices that expresses the problems that the world faces today and discovers a plethora of issues that speak of the need to assert a green paradigm[81] free of the stamp of lucrative colonial trade.

Fighting against Dams in the Name of Greater Common Good

In *Capitalism: A Ghost Story*, Roy criticizes the building of dams in the name of Common Good. She wonders, "How can they stop a dam?,"[82] referring to the Kalpasagar project[83] that has ambitious plans of supplying water to Special Investment Region (SIR) and Special Economic Zones (SEZ) projects that Roy labels as a self-governed corporate dystopia of "industrial parks, town-ships and mega-cities."[84,85] In an earlier political essay-manifesto titled "Greater Common Good," Roy defends the *Adivasi* tribal people,[86] who lost their homes and livelihood to the construction of the Sardar Sarovar Dam in the Narmada valley.[87] This essay's preface is dedicated to the River "Narmada and all the life she sustains."[88,89] The opening lines—"If you should suffer you should suffer in the interests of the country"[90]—are a nod to a speech made by India's first prime minister Jawaharlal Nehru[91] who described dams as temples of modern India while addressing homeless villagers who had been displaced by the Hirakud dam construction in 1948.[92] She reminds readers that displacement and evacuation of villagers will be inevitable once again with the Kalpasagar dam. Her essays aim to "puncture the myth about the inefficient, bumbling, corrupt, but ultimately genial, essentially democratic, Indian state."[93] In the power struggle between the protagonists of Common Good who support development ventures and adversaries who favor a pre-industrial dream, she accuses both groups of resorting to "deceit, lies, false promises and increasingly successful propaganda"[94] to configure a sense of false legitimacy. She implies that the "Iron Triangle" composed of politicians, bureaucrats, and dams work hand in glove with British consultants of the world to devise Environmental Impact Assessments that mask and hide the unavailability of water statistics, the destruction of flora and fauna, and the mass exodus of uneducated villagers. Roy accuses the Indian government of violating the human rights to a normal standard of life of innocent people as they "stand to lose their homes, their livelihoods, their gods and their histories."[95] She states that this quest for modernity, as demonstrated in the creation of the Dholera SIR,[96] where there is the risk of extinction of rare fish species, is typical of the selfish human decision to survive at the expense of wildlife. She insists on the need to find means of "exploiting nature while minimizing non-human claims to a shared earth."[97],[98] Roy's essay, with its hidden agenda of social and environmental advocacy, is imaginative and serves as "a catalyst for social action and [. . .] a full-fledged form of engaged cultural critique."[99] With time running out, Roy insists that words can prove to be the best arms to protect India's chemical polluted rivers. Her fight against the dam does not involve just the tryst with the Iron Triangle, but also encompasses the struggle to preserve a whole ecosystem of cropping and breeding patterns of humans and animal species. Through Roy's account the reader discovers a clear lay of the land defining "what happened where and when and to whom."[100] She alerts readers that the Narmada valley containing fossils, microliths of the Stone Age, and the history of the *Adivasis* was doomed in the name of Common Good.

CONCLUSION: MAKING THE MATERIAL TURN

This pedagogical approach considers the theoretical framework of the material turn which conceives matter as an agentic force with an effective and transformative power over human and nonhuman environments. Amitav Ghosh and Arundhati Roy explore the problems of conserving environmental biodiversity, distrust materialistic intent that crushes ordinary people, and take part in advocating the new emerging dynamics of making a material turn. Thus their texts offer possible ways of analyzing language and reality, human and nonhuman life, mind and matter, without falling into dichotomous patterns of thinking. Both authors probe into the reasons for the postcolonial writer's imaginative failure in the face of global warming. They insist on the need for the younger generation to take part in environmental politics and posit that this should be reflected in the literature of our age, in historical accounts, and in political decision-making of contemporary postcolonial governments. Ghosh and Roy resist stasis[101] and campaign for change by advocating new contemporary modes of thinking and imagination. This pedagogical approach of analyzing the material turn will enable students and teachers to contemplate how Ghosh's and Roy's ecocritical writing is the best of all cultural forms to voice ecocritical awareness.

Eco-narrative versus Eco-materialism

Ghosh's narrative in the Ibis Trilogy strives to remember the materialist colonial past, but at the same time criticizes contemporary geopolitics. He highlights the fact that the Indian Ocean welcomes sailors from India, China, Mauritius, Europe, and the United States, but the trade was colonial with the terms being dictated by the British Empire. Even today, the ocean remains the lieu of maritime trade affecting politics on land, agricultural production, and environmental policy making. This archive of unfair trade is evident in the IOR-ARC treaty signed by countries that share the Indian Ocean. Though created with the intention of being a platform for the peoples of the Indian Ocean Region to reconnect with each other, to discover their common heritage and deep-rooted affinities, to celebrate their shared cultural history and chart their own destinies, the free trade association has been criticized for having pitched "too high" or "too low" its tariffs and customs barriers. In India, the cultivation of opium poppies is now regulated with farmers producing mainly for medicinal or research purposes. Governmental proposals to create a drug-free state at the expense of poor farmers have resulted in these farmers losing their livelihood. He insists that mere eco-friendly policy labeling will not solve the world's problems. He calls on writers to recall and to write on issues that affect the world and to consider new eco-narratives that speak of the need to preserve life, encourage environmental advocacy, and bring ecocriticism closer to the material turn by highlighting how narratives and stories contribute to making meaning of the material forces and substance that rule the world. Ghosh's ecocritical writing shows that ignoring climate change and environmental hazards will certainly make future generations contemplate why their predecessors encouraged this time of Great Derangement.

Roy refers to the disparity between the rich and the poor as "this confederation of loyal, corrupt, authoritarian governments in poorer countries to push through unpopular reforms and quell mutines."[102] Through an enumeration of the TATA empire's material influence on our daily life, Roy contends that the ordinary man is "under siege":

> We all watch Tata Sky, we surf the net with Tata Photon, we ride in Tata taxis, we stay in Tata Hotels, sip our Tata tea in Tata bone china and stir it with teaspoons made of Tata Steel. We buy Tata books in Tata bookshops. *Hum Tata ka namak khatey hain.*[103] We're under siege.[104]

The essay-manifesto is the ideal mode for Roy's ecocritical enterprise of underlining the need for social and political change. It speaks of Roy's envisioning of a "postcolonial green" that campaigns for the transference from "red" to "green" politics and the need to take the material turn and dwell as responsible inhabitants who believe in global justice and sustainability on our planet Earth. Roy posits that the world needs a "new kind of politics. Not the politics of governance, but the politics of resistance. . . . The politics of forcing accountability."[105] She advocates writing that contributes to climate education and encourages the youth to join hands across the world and prevent the destruction of the planet Earth.

Amitav Ghosh advances the idea that a writer's imagination plays a vital role in shaping the minds of young readers. He states that "fiction, for one, comes to be reimagined in such a way that it becomes a form of bearing witness, of testifying, and of charting the career of the conscience."[106] Through her writing, a form of nonviolent dissent to change the world, Arundhati Roy rightly advocates that "the only thing worth globalizing is dissent. It's India's best export."[107] The pedagogical approach of analyzing the climate change and sustainability challenges of the century through the lens of making a material turn in fiction and nonfiction is the ideal mode to capture the attention and interest of the youth at large.

NOTES

1. Amitav Ghosh has a doctorate in social anthropology from the University of Oxford. His works reveal his study of the ways in which people live in different social and cultural settings across the globe. His other works of fiction are *The Circle of Reason* (1986); *The Shadow Lines* (1988); *In an Antique Land* (1992); *The Calcutta Chromosome: A Novel of Fevers, Delirium, and Discovery* (1995); *The Glass Palace* (2000); *The Hungry Tide* (2004); the Ibis Trilogy comprising *Sea of Poppies* (2009), *River of Smoke* (2011), *Flood of Fire* (2015); and *Gun Island* (2019).

2. Arundhati Roy (born November 24, 1961) is an Indian author, actress, and political activist-writer who was awarded the Booker Prize in 1997 for her semi-autobiographical-debut novel *The God of Small Things*. This chapter will highlight her involvement in environmental and human rights causes.

3. Ghosh's other works of nonfiction include *Dancing in Cambodia and at Large in Burma* (1998), *The Imam and the Indian* (2002), and *Incendiary Circumstances: A Chronicle of the Turmoil of Our Times* (2005).

4. Amitav Ghosh, *The Great Derangement: Climate Change and the Unthinkable* (Chicago, IL: The University of Chicago Press, 2016), 10.

5. Ghosh, *The Great Derangement: Climate Change and the Unthinkable*, 11.

6. Ghosh, 11.

7. Arundhati Roy's political essays have been compiled into the following collections: *The End of Imagination* (1998), *The Algebra of Infinite Justice* (2001), *Power Politics* (2002), *War Talk* (2003), *An Ordinary Person's Guide to the Empire* (2005), *The Shape of the Beast* (2008), *Listening to Grasshoppers: Field Notes on Democracy* (2009), *Broken Republic* (2011), and *My Seditious Heart* (2019).

8. In a legal affidavit entitled "On Citizens' Rights to Express Dissent" in her work *My Seditious Heart*, Roy quotes an extract from the Supreme Court Order dated

October 15, 1999 that accuses Roy's writing of violating the dignity of the court and polluting the stream of justice.

9. Arundhati Roy, "On Citizens' Rights to Express Dissent," in *My Seditious Heart: Collected Nonfiction* (Toronto: Hamish Hamilton, 2019), 156.

10. Arundhati Roy has received awards for her essay writing, like the Lannan Award for Cultural Freedom in 2002, the 2003 Noam Chomsky Award, the Sydney Peace Prize in 2004, the 2005 Sahitya Kademi Award, which she declined, and the Norman Mailer Prize in 2009 for Distinguished Writing.

11. Serenella Iovino and Serpil Oppermann, *Material Ecocriticism* (Indiana University Press, 2014), 1–2.

12. Ghosh, *The Great Derangement: Climate Change and the Unthinkable*, 8.

13. Ghosh, 9.

14. Arundhati Roy, "Arthur Miller Freedom to Write Lecture," *The Guardian*, May 13, 2019, pars. 2–5. https://www.theguardian.com/commentisfree/2019/may/13/arundhati-roy-literature-shelter-pen-america.

15. Roy, "Arthur Miller Freedom to Write Lecture," par. 8.

16. Roy, "Arthur Miller Freedom to Write Lecture," par. 8.

17. Ghosh holds a PhD in Social Anthropology from the University of Oxford. He also received honorary doctorates from Queen's College in New York and the Sorbonne.

18. Roy has a degree in architecture from the Delhi School of Architecture. She is also a screenplay writer and has wrote the screenplays for *In Which Annie Gives It Those Ones* (1989), a film based on her experience as a student of architecture.

19. Roy criticizes the dominance of richer nations over developing countries by means of unequal conditions of economic exchange. She claims that neo-imperialism results in richer nations restricting poorer ones from stepping out of the roles that the former had defined for them, like, for instance, reducing the latter to providers of raw materials and cheap labor.

20. https://www.amitavghosh.com/; https://www.weroy.org/.

21. Ghosh, *The Great Derangement: Climate Change and the Unthinkable*, 15–17.

22. The meandering River Padma is one of the major rivers of Bangladesh. Evidence from satellite imagery reveals that it has been constantly gaining volume and changing its trajectory during the past decade. The river's source is at the junction of the Ganges and Jamuna Rivers in India. It then merges with the Meghna River in Bangladesh and ultimately empties into the Bay of Bengal.

23. The Ganges River (Ganga in Hindi) is the great river of the plains of Northern Indian subcontinent. It rises in the Himalayas and empties into the Bay of Bengal. As it flows through the Indo-Gangetic Plain, it irrigates one-fourth of the country, and has been the cradle of successive civilizations from the Mauryan Empire of Ashoka in the third century BC to the Mughal Empire, founded in the sixteenth century.

24. In *The Hungry Tide*, Ghosh speculates on the origin of the name of the mangrove forests and the anthropological, botanical, geo-tidal, and historical influences in the bearing of its name. "Sundarbans" means "beautiful forest."

25. Ghosh, *The Great Derangement: Climate Change and the Unthinkable*, 6.

26. *Bhatir desh* means "land of the low tide" or "tidal country." Ghosh explains ". . . in the record books of the Mughal emperors this region is named not in reference to a tree but to a tide—bhati. And to the inhabitants of the islands this land is known as *bhatir desh*—the tide country—except that bhati is not just the "tide" but one tide in particular, the ebbtide, the bhata: it is only in falling that the water gives birth to the forest. To look upon this strange parturition, midwived by the moon, is to know why the name "tide country" is not just right but necessary" (Ghosh, 2004, 8).

27. Amitav Ghosh, *The Hungry Tide* (New Delhi: Ravi Dayal, 2004), 7.

28. Hugli River, also spelled Hooghly, is a river in West Bengal state, in Northeastern India. It is an arm of the Ganges (Ganga) River. It branches off the Ganges and provides access to Kolkata (Calcutta) from the Bay of Bengal.

29. The Meghna River is the major watercourse of the Ganges in Bangladesh. It receives the combined waters of the Padma and Jamuna (the name of the Brahmaputra in Bangladesh) rivers near Chandpur in Bangladesh.

30. Ghosh, *The Hungry Tide*, 246.

31. Bon Bibi or Ban Bibi is a reference to the guardian spirit of the Sundarbans who protects woodcutters, inhabitants, and travelers from tiger attacks.

32. Dokhin Rai or Dakshin Rai is a revered deity in the Sundarbans who rules over beasts and demons. He is the arch enemy of Bon Bibi.

33. After a twenty-year gap of novelistic silence, Roy creates a fictional space in *The Ministry of Utmost Happiness* that reveals the lives of hijras (eunuchs) who live in communities that have created alternative structures of kinship, resistance, and romance.

34. Arundhati Roy, *The Ministry of Utmost Happiness* (New Delhi: Penguin, 2017), 195.

35. Roy Arundhati, "Power Politics," in *The Algebra of Infinite Justice* (New Delhi: Penguin, 2002), 147.

36. Arundhati Roy, *Capitalism: A Ghost Story* (London: Verso, 2015), 13.

37. Roy, *Capitalism: A Ghost Story*, 2.

38. Ironically Vedanta's tagline is "Mining Happiness." Its British-based parent company Vedanta Resources, a natural resources conglomerate, aims at creating long-term shareholder value through research, discovery, acquisition, sustainable development, and utilization of natural resources. The company mines natural resources, but Roy muses on whether they really mine happiness.

39. In 2014, the Jindal Group Jindal was under investigation in connection with the allocation of coal mining rights in a scandal called "Coalgate" by India's media.

40. The Pollution Control Board of the State of Madhya Pradesh has imposed a fine of 50 lakh rupees (approximately $70,275) toward farmers' compensation. The Essar plant was responsible for crop damage caused by overflowing ash water from one of its power plants in Singrauli.

41. Think Festivals were organized during the period 2011 to 2013 and aimed at creating thought-provoking, egalitarian ideas from across the globe.

42. It was begun in 1907 by Jamshedji Tata, India's pioneer industrialist. It is today one of the world's most geographically diversified steel producers with operations and commercial presence across the world.

43. The Anglo-Australian multinational is engaged in diamond mining in the Indian state of Madhya Pradesh.

44. Jaipur Literary Festival is an annual literary festival held in the Indian city of Jaipur. It is a five-day festival where writers, thinkers, politicians, and entertainers meet on one stage to engage in thoughtful debate and dialogue.

45. Salman Rushdie, "Is Nothing Sacred?," in *Imaginary Homelands: Essays and Criticism, 1981–91* (London: Vintage Books, 2010), 428.

46. In his essay "Is Nothing Sacred?" Salman Rushdie leads the reader into the unassuming little room of literature tucked away in a corner of the large rambling house of world activity. The room is alive with ideas and dialogues. Rushdie articulates the dire need to preserve this privileged arena of creative enterprise and includes readers in the enterprise of assuring the survival of the little unassuming room of literature by putting forth the idea that creating a literary masterpiece is not an affair of individual creative genius, but a joint endeavor.

47. Roy, *Capitalism: A Ghost Story*, 19.

48. An acronym coined by Jim O'Neill of Goldman Sachs in 2001. The BRIC countries comprising of Brazil, Russia, India, and China are not a political alliance like the European Union or a formal trading association. They assert their power as an economic bloc by signing formal trade agreements together and attending summits together. In 2010, South Africa joined this group and they now concert with each other's interests as BRICS countries. By 2050, these countries will probably be wealthier than most of the current economic powers.

49. In the 1950s, the name "the Golden Triangle" was given to the area of Mainland Southeast Asia where most of the world's illicit opium originated. This name "Golden Triangle" was first coined by the U.S. vice-secretary of State, Marshall Green, during a press conference on July 12, 1971.

50. Pierre-Arnaud Chouvy, *Opium: Uncovering the Politics of the Poppy* (Cambridge, MA: Harvard University Press, 2010), 23.

51. Amitav Ghosh, *Sea of Poppies* (London: John Murray, 2008), 12.

52. The Andaman Sea takes its name from the Andaman Islands, a union territory of India. It is situated in the Northeastern Indian Ocean. It is the sea link between Myanmar and other South Asian countries. It is on the shipping route between India and China, via the Strait of Malacca.

53. A reference to the old foreign enclave in Canton (Guangzhou) which was also known as the "Thirteen Factories" (Saap Sam Hong). Fanqui-town does not exist today as these factories were burnt down in 1856 and were never rebuilt again.

54. The novel's diegesis reveals how the British imposed their trade in Indian opium upon China through the two so-called Opium Wars (1839–1842 and 1856–1860), thus not only monopolizing the delivery of the narcotic to the European market, but also changing the course of economic and political relations between the East and the West.

55. Carl A. Trocki, *Opium, Empire, and the Global Political Economy: A Study of the Asian Opium Trade 1750–1950* (London: Routledge, 1999), 7–8.

56. Julie Rivkin, Michael Ryan, and Edward Said, "Orientalism," in *Literary Theory, an Anthology* (Malden, MA: Blackwell, 1998), 881.

57. Edward Said in his work Orientalism describes "Orientalism" as the Western attitude that views Eastern societies as exotic, primitive, and inferior.

58. It is interesting to make a reference to *Opium: Uncovering the Politics of the Poppy*, where Pierre-Arnaud Chouvy explains how a triangular trade developed between Britain, India, and China in which Indian opium provided the silver required to buy tea legally from China for shipments to London. He speculates on how opium ensured the profitability of colonial trade by enabling the British to balance their trade deficit with China: "Fearing that payment for Chinese imported goods (tea consumption was growing fast in Britain when only China produced the leaves) would deplete their silver reserves, the British resorted to opium, a product of their Indian colony, as a means of payment." It is interesting to know that through the opium trade of the English East India Company and the tax on Malwa opium, the British made a profit of $15,488,000.

59. Nemesis is the first British oceangoing iron warship. It was launched in 1839 by the British East India Company to take part in the Opium Wars. It was also known as "The Devil Ship" due to the havoc it stirred up during attacks.

60. Amitav Ghosh, *Flood of Fire* (London: John Murray, 2016), 509–10.

61. Ghosh, *Flood of Fire*, 505.

62. The phrase "The Empire Writes Back to the Centre" was originally used by Salman Rushdie in his article "The Empire Writes Back with Vengeance." Rushdie's pun is also from the film *Star Wars: The Empire Strikes Back*. In his article, Rushdie referred to the erstwhile British colonies as the Empire as they regained independence and wrote back to the former colonizer. Many diasporic writers in the UK have used various literary strategies of decolonization to set the record straight by writing back to the Centre or the former British Raj. Australian critics, Bill Ashcroft, Gareth Griffiths, and Helen Tiffin, entitled their work *The Empire Writes Back* to highlight the powerful forces acting on language in the postcolonial text and show how these texts constitute a radical critique of Eurocentric notions of literature and language.

63. Bill Ashcroft, Gareth Griffiths, and Helen Tiffin, *The Empire Writes Back: Theory and Practice in Post-Colonial Literatures* (Brantford, ON: W. Ross MacDonald School Resource Services, 2011), 221.

64. Ghosh, *Flood of Fire*, 484.

65. The area of Southeast Asia encompassing parts of Burma, Laos, and Thailand, significant as a major source of opium and heroin.

66. The name "Golden Crescent" is of unknown authorship and is similar to that of "Golden Triangle and can be considered as the Southeast Asian alter ego. 'Its "Crescent" refers to the Muslim dimension of this opium-producing region comprising the countries of Afghanistan, Iran, and Pakistan . . . Afghan and Pakistani opium was first marketed for Persian and Indian consumers" (Chouvy 28–9). So, the name "Golden" refers to the Muslim dimension of this opium-producing region comprising the aforementioned countries. Interestingly, Afghan and Pakistani opium was first marketed for Persian and Indian consumers.

67. Roy, *Capitalism: A Ghost Story*, 8.

68. Roy, *Capitalism: A Ghost Story*, 7.

69. Roy, *Capitalism: A Ghost Story*, 9.

70. Roy, *Capitalism: A Ghost Story*, 9.

71. Roy, *Capitalism: A Ghost Story*, 9.

72. Roy, *Capitalism: A Ghost Story*, 9.

73. Roy, "Power Politics," 150–1.

74. Roy, "Power Politics," 152.

75. Three mineral-rich Indian states mineral-rich Orissa where hundreds of indigenous tribespeople are battling to stop London-listed Vedanta Resources Plc from extracting bauxite from what they say is their sacred mountain.

76. Roy, *Capitalism: a Ghost Story*, 11–12.

77. A Latin term from the history of Pharmacy. It is the branch of medical science that deals with the sources, nature of plants, properties, and preparation of drugs.

78. The dried young leaves of camellia sinensis have been used since ancient times to make green tea. It contains a high concentration of antioxidants known as polyphenols as the leaves are not fermented.

79. Amitav Ghosh, *River of Smoke* (London: John Murray, 2011), 39.

80. Ghosh, *River of Smoke*, 82–3.

81. The green paradigm refers to the quest of embracing social and ecological values that protect the planet.

82. Roy, *Capitalism: A Ghost Story*, 15.

83. The building of Kalpasar dam in Gujarat is still underway. Once completed, it will be thirty-four-kilometers long stretching across the Gulf of Khambat. There are plans to configure a ten-lane highway and a railway line on top of it. This sweet water reservoir of Gujarat's rivers has a further confluence and network of 168 dams that are mostly privatized and have been planned across war-hit and tension-prone zones of Kashmir and Manipur.

84. Special Investment Region (SIR) refers to an investment region of 100 square kilometers in Gujarat region of India. The aim is to set up world-class hubs of economic activity. SEZ refers to Special Economic Zones where government agencies including private companies may be assigned powers and functions to promote the development of a SIR.

85. Roy, *Capitalism: A Ghost Story*, 16.

86. The *Adivasis* are the earliest inhabitants of the Indian subcontinent, who are considered as the indigenous tribes who occupied the hill and mountainous regions of India. The term *Adivasi* was coined in the 1930s to give a sense of identity to the various indigenous tribes of India. In Hindi *adi* means "of earliest times" and *vasi* means "inhabitant."

87. The Sardar Sarovar Project across the Narmada is being built at an estimated cost of Rupees 392.4 billion (approximately 8 billion USD). It is the world's second largest concrete gravity dam, with the world's third highest spillway discharging capacity of 87,000 cubic meters, and will be the largest irrigation canal in the world after its completion, irrigating Kutch and Saurashtra. The government of India calls it an "eco-friendly" indigenous hydropower reservoir.

88. Roy Arundhati, "Greater Common Good," in *The Algebra of Infinite Justice* (New Delhi: Penguin, 2002), 46.

89. The river Narmada is the fifth longest river in the Indian subcontinent and flows through Central India, through the state of Madhya Pradesh. Thirty mega-dams are being constructed along the river basin, which is home to teak forests.

90. Roy, "Greater Common Good," 47.

91. The ironic reference to suffering for the cause of common good mentioned in Nehru's "dam" speech has inspired the title of the essay "Greater Common Good."

92. The Hirakud dam is India's first river valley project after India's independence (August 14, 1947). It is India's largest dam and the world's fourth largest barrage. It provides hydroelectricity to villages along the Mahanadi river.

93. Roy, "Greater Common Good," 69.

94. Roy, "Greater Common Good," 52.

95. Roy, "Power Politics," 155.

96. The Dholera Special Investment Region covers approximately 920 square kilometers and covers 22 villages that are strategically situated between the industrial zones of Ahmedabad, Baroda, Rajkot, and Bhavnagar in Gujarat State. This industrial hub will have a six-lane access-controlled expressway, metro rails, and an international airport.

97. Graham Huggan, *Postcolonial Ecocriticism: Literature, Animals, Environment* (London: Routledge, 2015), 5.

98. Graham Huggan and Helen Tiffin in *Postcolonial Ecocriticism* examine the relationship between humans, animals, and the environment in postcolonial literary texts. They hold that human societies need to consider their relationship with nonhuman species with whom they share the planet. They insist on the need to imagine new ways of creating awareness of these ecologically connected groupings.

99. Huggan, *Postcolonial Ecocriticism*, 12

100. Roy, "Greater Common Good," 109.

101. This echoes Rushdie's preference for an aesthetics of inconstancy in writing. He claims, "Stasis, the dream of eternity, of a fixed order in human affairs, is the favoured myth of tyrants; metamorphosis, the knowledge that nothing holds its form, is the driving force of art" (Imaginary 291).

102. Arundhati Roy, "Confronting Empire," in *My Seditious Heart: Collected Nonfiction* (Toronto: Hamish Hamilton, 2019), 223.

103. Meaning 'we eat TATA salt' in Hindi.

104. Roy, *Capitalism: A Ghost Story*, 20

105. Roy Arundhati, "The Ladies Have Feelings," in *The Algebra of Infinite Justice* (London: Flamingo, 2002), 215.

106. Ghosh, *The Great Derangement: Climate Change and the Unthinkable*, 128.

107. Roy, "The Ladies have Feelings," 215.

REFERENCES

Ashcroft, Bill, Gareth Griffiths, and Helen Tiffin. *The Empire Writes Back: Theory and Practice in Post-Colonial Literatures*. Brantford, ON: W. Ross MacDonald School Resource Services, 2011.

Chouvy, Pierre-Arnaud. *Opium: Uncovering the Politics of the Poppy*. Cambridge, MA: Harvard University Press, 2010.

Ghosh, Amitav. *The Hungry Tide*. New Delhi: Ravi Dayal, 2004.

———. *Sea of Poppies*. London: John Murray, 2008.

———. *River of Smoke*. London: John Murray, 2011.

———. *Flood of Fire*. London: John Murray, 2016.

———. *The Great Derangement: Climate Change and the Unthinkable*. Chicago, IL: The University of Chicago Press, 2016.

Huggan, Graham. *Postcolonial Ecocriticism: Literature, Animals, Environment*. London: Routledge, 2015.

Iovino, Serenella, and Serpil Oppermann. *Material Ecocriticism*. Indiana University Press, 2014.

Rivkin, Julie, Michael Ryan, and Edward Said. "Orientalism." Essay. In *Literary Theory, an Anthology*, 873–885. Malden, MA: Blackwell, 1998.

Roy, Arundhati. "Greater Common Good." In *The Algebra of Infinite Justice*, 46–141. New Delhi: Penguin, 2002.

———. "Power Politics." In *The Algebra of Infinite Justice*, 143–184. New Delhi: Penguin, 2002.

———. "The Ladies Have Feelings." In *The Algebra of Infinite Justice*, 187–215. London: Flamingo, 2002.

———. *Capitalism: A Ghost Story*. London: Verso, 2015.

———. *The Ministry of Utmost Happiness*. New Delhi : Penguin, 2017.

———. "Arthur Miller Freedom to Write Lecture." *The Guardian*, May 13, 2019, https://www.theguardian.com/commentisfree/2019/may/13/arundhati-roy-literature-shelter-pen-america.

———. "Confronting Empire." Essay. In *My Seditious Heart: Collected Nonfiction*, 221–226. Toronto: Hamish Hamilton, 2019.

———. "On Citizens' Rights to Express Dissent." Essay. In *My Seditious Heart: Collected Nonfiction*, 150–159. Toronto: Hamish Hamilton, 2019.

Rushdie, Salman. "Is Nothing Sacred?" Essay. In *Imaginary Homelands: Essays and Criticism, 1981-91*, 415–429. London: Vintage Books, 2010.

Trocki, Carl A. *Opium, Empire, and the Global Political Economy: A Study of the Asian Opium Trade 1750-1950*. London: Routledge, 1999.

Chapter 3

Climate Crisis Confluence, History, and Social Justice

How Race, Place, Privilege, Past, and Present Flow Together in YA Literature

Anna Bernstein and Kaela Sweeney

Ibram X. Kendi wonders "How many Americans ask, 'Do you think racism is still a problem?' or 'Do you believe the globe is warming?' as if society should value ignorance in the face of scientific certainty."[1] He notes the overlap of systemic denial of racist structures and avoidance of the climate crisis as ways in which Americans reject scientifically observational data in favor of "belief." In the American public school system, the divide between belief and fact becomes blurred when topics arise that have been politicized and polarized. The topics of racism and climate change inherently permeate each subject area but are often left undiscussed and unexamined. The confluence of race, place, and privilege can be seen in discussions of climate change as teachers attempt to bridge the gap between what students are told to believe and what science presents as believable. This is where the genre of young adult (YA) climate fiction becomes a tool for dismantling oppressive, widely held views.

In order to delve more deeply into Kendi's ideas about the denial of racism and impending climate crisis, it is imperative to understand the definitions of terms that are used throughout discussions of each. When we talk to young people about race, we want to compel a dual understanding: race is both a construct and an unavoidable reality. We want students to hold the space for a rejection of race as a determinable reality while also vehemently affirming the experiences of racism that are so prominent in the American experience. Dorothy Roberts notes that "race applied to human beings is a political division: it is a system of governing people that classifies them into a social hierarchy based on invented biological demarcations."[2] In this way, educators can frame race as a tool of oppression rather than an observable truth. But at the same time, we must assert to students that despite the invalidity of race, racism is an observable and measurable occurrence, one that they in turn have the power to confront and dismantle.

Intimately tied to race is the idea of place, the physical and cultural centers of existence for humans navigating racism and the climate crisis. Place refers to both the geographic location of a people and the position of a society in the greater global milieu. The intersection of race and place is particularly important in the discussion of climate change in that the negative effects of the climate crisis are disproportionately observable in both the Global South and among communities of color. When investigating how climate change affects communities, it is impossible to separate racial disparity from environmental justice. In YA fiction, climate change reflects reality through the setting of each story and the choice of narrator or protagonist. Our particular context involved a public school that is predominantly white due to the impact of gentrification. In order to highlight the parallel between our own community's issues of inequity, we chose texts that parallel the lived experiences of the global majority in settings that are particularly vulnerable to climate change.

There is no discussion of race or place without the underlying foundation of privilege that supports both in the perpetuation of structural oppression. In *So You Want to Talk about Race*, Ijeoma Oluo defines privilege as "an advantage or a set of advantages that you have that others do not."[3] With this meaning in mind, the exploration of race, place, and privilege through the lens of climate change becomes a discussion about the difference between responsibility and repercussions. Privilege in the conversation about climate crisis refers to the ability of postindustrial nations to produce the majority of climate change catalysts while avoiding the most detrimental effects of environmental consequences and outcomes. Privilege enables mostly white and socioeconomically elevated countries to deny the reality of climate change because they are less impacted than the nations facing discernible detriment in the form of increasing catastrophic weather events or further depletion of natural resources.

While NASA defines climate change as "long-term change in the average weather patterns that have come to define Earth's local, regional and global climates," we will refer to the intersection of race, place, privilege, and climate as the "climate crisis." As educators, we choose to use this phrase to emphasize the immediacy of human involvement in the physical degradation of the global environment; it also highlights the importance of investigating causes and taking tangible, measurable actions to reverse negative effects. The etymology of the word "crisis" traces its roots back to the Greek word that means "to decide" which aligns perfectly with the philosophical and scientific notion that humans have made decisions that unquestionably led to a severity in climate change. Importantly, it also implies that those same humans have the power to make environmentally dynamic decisions. The phrase "climate crisis" thus suggests that humans have the culpability of decision-making both in contributing to and mitigating the impacts of climate change.

Race, place, and privilege converge in the dystopian climate fiction of YA literature as the inseparable factors that generate and perpetuate the climate crisis. With young narrators or protagonists, YA climate fiction empowers a group of people with the least guilt but the most to lose in the future trajectory of climate change. Utilizing these literary examples of youth in action allows students to envision future consequences while imagining and proposing possible solutions of remediation. This imagining comes in the form of storytelling and future creation. Storytelling is the natural human process of sensemaking and solution processing. The act of narration places power in the hands of the narrator as they decide how the story will go, how it will be told, and with whom it will be shared. In this way storytelling is not only an act of creation, but a willful expression of re-creation and future molding. Allowing young people to see themselves in imagined potential situations of dire climatic circumstance forces circumspection around issues spotlighted in the texts. This important connection encourages creative problem-solving and imaginative fabrication of solutions. Young people can use storytelling to reimagine the worlds they inherit. This reimagining can include working through possible interventions and crafting future worlds with those possible measures in place, but within the safe confines of literature. This kind of experimentation empowers youth to position environmental justice at the forefront of policy and conversation, without the immediate danger of the climate crisis' physical ramifications.

Fiction texts expose readers to scientific evidence in a subversive way, often inserted by the author in a way that feels authentic and natural rather than politicized indoctrination. Fiction may help the stress of climate crisis reality by providing readers with the same educative information without the overpowering sense of dread that can come with nonfiction texts of more tedious data. Readers can experience the dire nature of the crisis while still being offered hope in an imaginative and malleable genre. When discussing how people react when confronted by pure science and reason Ibram X. Kendi says that "in disbelieving the present observable realities, they certainly disbelieve future projections"[4] which highlights the failure of a fully nonfiction approach to convey the immediacy of the climate situation and persuade the general public about its importance. Storytelling with individual focus allows the climate crisis to become personalized and poignant for YAs, supplying them with the space to envision a future at the intersection of race, place, and privilege that either suffers from or overcomes environmental destruction. In this chapter we will present three core YA climate fiction texts that grapple with these important elements as well as nonfiction sources that provide foundational, science-based reasoning to understand the experiences in the novels. Pairing these resources in an instructional setting fosters a bridge to an understanding of the climate crisis as well as a call to action for students.

CLIMATE FICTION LITERATURE CIRCLES: PURPOSE

The ecology of our eighth grade English Language Arts curriculum features a pattern of how change happens within and among communities. Our school district's scope and sequence offer essential questions for quarterly units starting with "How do genre and text structure impact and contribute to theme and meaning? How do the authors force readers to reflect on their perceptions of culture, and how do those reflections shape identity?" The units in the second half of the year explore the following questions: "How do ideas of freedom and equality change when race, culture, or gender are involved? How does a 'single story' about a group of people affect all of humanity?" The district's final essential question for eighth grade English Language Arts is "How do people use their understanding of who they are to become agents for change?" These questions explore genre, identity, and activism, which are all essential to our study of climate fiction. In addition to our school district's essential questions, eighth graders complete their International Baccalaureate Middle Years Programme Community Project in English Language Arts class. Through their work on this service project, students address specific problems and needs within their communities and develop plans of action to address those needs. Although social justice is not the sole thread between units, our students tend to focus on the theme of justice as they identify problems within their personal communities. Studying climate fiction allows students to use the skills they practiced in previous units while also seeing real and imagined examples of young people taking action in their communities.

The climate fiction study serves as our culminating unit for several reasons: (1) Many students are already interested in the climate crisis. Throughout the school year, our students work on community action projects based on problems they identify within their communities. Last year, many of our students chose environmentally focused issues. In fact, the climate crisis is something with which young people all over our country and globe are forced to consider. Those students who are initially interested are often engaged with the dystopian YA titles that serve as anchor texts in our literature circles. Students get to choose the book they study and discuss in literature circles. (2) Conversations around environmental inequities, climate change, and environmental justice are necessary in the secondary classroom. According to the 2019 Resolution on Literacy Teaching on Climate Change, the National

Council of Teachers of English commits to the idea that "understanding climate change challenges the imagination; addressing climate change demands all the tools of language and communication, including the ability to tell compelling stories about the people and conflicts at the heart of this global discussion."[5] As such, (3) we believe reading and analyzing various forms of environmental writing empowers our students to explore how they affect and are affected by the environment as well as the interconnected communities and systems around them. A study of the climate crisis through literature is a study of converging disasters: "Climate change is an accelerator that exacerbates economic, racial, and social inequality. English language arts involve understanding and creating relationships with and between people and characters mediated through language, texts, and media."[6] Learning about climate change and environmental justice necessitates action at the convergence of our students' various spheres of influence. At this intersection, students can use literacy as a tool for change in their communities. It is through climate fiction, which offers a narrative of converging crises, that we began our study of three texts that merge at the junction of race, place, and privilege.[7]

Like many middle and secondary schools in the United States, our community sits at this intersection of race, place, and privilege. As our students work on their eighth grade community action projects, they often struggle to see the impact of various systemic issues on the problems they identify. Their focus frequently lands on the impact of individuals rather than systems. With this in mind, we set out on a journey of selecting stories that explore this intersection that societal systems create. For our unit, we asked students to choose one of the three anchor texts to discuss in small groups. We then engaged in whole-class discussions where students reading different texts were able to act as experts and share ideas that may offer new perspectives for other literature circle groups. As we thought of authentic learning tasks that require students to think about systems rather than individuals, we reflected on the different ways of reading we expected from our students throughout this unit. We planned discussions and paired texts that guided students through an exploration of how race, place, and privilege intersect in climate fiction and environmental storytelling. As a result, each YA climate fiction text study in this chapter is divided into three sections that guided our collective study of the texts: (1) Reading to Make Sense of the Climate Crisis, (2) Reading to Explore Climate Crisis "Form," and (3) Pairing Texts to Strengthen Our Understanding.

YA Climate Fiction Text Study #1: *Dry* by Jarrod Shusterman and Neal Shusterman

Dry follows several young adults as they fight to survive a drought crisis in California known as the "Tap-Out." This example of climate fiction offers our students multiple "forms" to explore as they think about how storytelling can play a role in making sense of science and inspiring action within communities. The exploration of a water crisis in *Dry* presents a series of converging crises to consider: drought, loss of public services, privilege based on place, and climate migration. It also explores the role of community as the main characters grapple with how individuals and groups of people respond in moments of crisis. In the climate story that *Dry* presents, students see young people making sense of disaster amid the polarizing emotions experienced in adolescence. Alyssa, along with her brother, her neighbor, and other young people she encounters on her journey to find water and survive, makes decisions to help her family and community navigate the disastrous effects of the California drought. The characters in *Dry* survive merging disasters by managing grief and loss, working through differences in beliefs among community members, and taking action to create change.

Reading Dry *to Make Sense of the Climate Crisis*

Dry begins with a dedication to "all those struggling to undo the disastrous effects of climate change." The Shustermans name the disaster with intentional placement in the opening pages. Precise climate language and terminology was important for us as we set up the context for the book. However, beyond the introduction, climate language is not overtly mentioned, making the story accessible for students who may not have that background knowledge. Our intentional climate crisis lens creates a purpose for each literature group. The questions students explore immediately move beyond "What is the problem?" to "How does climate change disproportionately affect neighborhoods, cities, states, and countries that are not consumers of climate-change producers?" and "Whose voices are a part of the story and the solution? Whose voices are left out?" We encourage students to explore the issues of drought, water scarcity, inequitable access to resources, and climate migration as direct effects of the water crisis depicted in *Dry*.

Through careful intentional analysis of the novel's characters, our students notice how the varying identities and environments of each character create unique reactions and experiences when fighting for survival. These distinctions between characters highlight a truth that scientific data often reveals: those who are entrapped by the interconnecting crises stemming from environmental disaster are often not the ones responsible for causing the disaster.

Reading Dry *to Explore Climate Crisis "Form"*

Telling the story of converging crises presents a challenge, so we ask students to consider the form, structure, and techniques writers must consider when dealing with environmental disasters and community calls to action.

Essentially, we ask students to engage in the practice of reading as writers. We prompt students to think about how the writer of their literature circle group's text helps the reader comprehend the severity of a crisis while simultaneously offering glimpses into the humanity and familiar communities of the characters. Through the analysis of author's craft, we want students to both evaluate the effectiveness of stylistic choices and begin to use them in their own writing.

One of the first writing techniques we hope students will identify and analyze with their groups is the Shustermans' use of literary comparisons to give readers an explanation that can anchor those details that may be most difficult to imagine without lived experience. Alyssa's perspective of the "Tap-Out" begins with a kitchen faucet, a ubiquitous appliance that serves as the anchor for readers in this climate situation. However, in this moment of crisis, the kitchen faucet "coughs and wheezes like it's gone asthmatic. It gurgles like someone drowning. It spits once, and then goes silent."[8] Student groups notice that giving human characteristics to a sink helps them understand the severity of the situation. If the goal of climate fiction is to make environmental issues more accessible and to elicit support for collective activism, an analysis of these tools, like the personification of a sink, can serve as mentor sentences. They are textual moments students can use to hone their own climate writing skills. Additionally, students learn to evaluate the value of a writing tool for the intended goal of environmental storytelling. In turn, we hope that students will consider using a literary comparison of some sort as they write their own climate narratives and conduct interviews or community projects that will garner support and activism.

The groups that read *Dry* also analyze specific choices that focus on the use of multiple narrators and interspersed human-interest "snapshots" in order to provide the reader with varying perspectives of the climate crisis. The novel alternates between first-person perspectives of the characters, Alyssa, Kelton, Jacqui, and Henry. Each character's perspective highlights differing experiences based on identity, place, and privilege. However, when thinking about the challenging task of representing converging crises in literature, the Shustermans' use of the "snapshot" offers a unique "form" or writer's technique to examine with students in hopes that they can use it in their own writing and community action projects. While students can read about statistics that illustrate how environmental crises affect communities and individuals, they gain a better understanding of these impacts when they read these varying perspectives through snapshots. It is through the perspective of an activist in a snapshot that students see the reasoning behind "Raw. Tangible. Action."[9] And it is through the snapshot of an Orange County Water District employee that students get a glimpse into the fragility of utilities and government systems in moments of environmental crisis:

> With all the county's main water mains on emergency shutdown, and endless glitches in the computers trying to redirect what water is left, he's been transporting water manually to high-priority facilities. Just yesterday he drove one of a dozen tanker trucks delivering the contents of a high school swimming pool to Camp Pendleton Marine Base. But desperate times call for desperate measures, and water managers are scrambling to keep the sky from completely falling.[10]

The stand-alone "snapshot" causes students to acknowledge that the act of telling one climate story addressing race, place, and privilege is the act of telling multiple and varied stories concurrently. This snapshot form illustrates how two people may experience environmental disasters in vastly different ways. The short form causes students to ask questions such as "How might an activist experience this differently than a plant manager or a firefighter or an older woman trapped on the interstate?" Students get a glimpse into the inner thoughts of a news anchor, a National Guard pilot, and a Disneyland worker. These case studies model identity explorations to help students grasp the fact that intersecting identities change the way an individual navigates a given crisis. These fictional snapshots can be connected to real news events and people in similar situations. For example, students could read about the people and events that took place in February 2021 during the utility crisis in Texas or the water crisis in Jackson, Mississippi. Students could do their own research to understand how these crises impact different people in different ways based on their privilege and place. The fictional snapshots and journalistic representations of real people also offer templates for students as they work on sharing stories that incite action in their own communities.

Pairing Nonfiction Texts with Dry

For some students, recognition of the intersecting systems that cause the catastrophic damage of the "Tap-Out" occurs through the study of paired articles and videos. These pairings further open conversation about the role of storytelling in collective efforts to combat climate change. For example, *The New York Times* 2019 article, "'Turn Off the Sunshine': Why Shade Is a Mark of Privilege in Los Angeles," illustrates the cross section of income inequality and climate change's impact on shade equity based on race, place, and privilege.[11] Studying this article, students see the clear relationship between climate change and inequality. We asked groups to have discussions about what a city planned with environmental justice in mind might look like. In discussion forums, students noticed clear differences in neighborhoods, including access to green spaces based on privilege in their own communities. Other students realized yet another converging crisis with environmental justice and transportation when they pointed out that "even though someone could go drive somewhere (to downtown or a park), people who live in lower income areas tend to have less transportation, or working cars." Yet another student made the connection that support for communities is inequitable "especially during disasters." When given climate stories and connected nonfiction texts that highlight climate injustices, students begin to understand links between the equity issues presented by the texts and those that exist in their communities.

Additional articles that serve as exemplars to help students think about the relationship between equity and climate change include *The New York Times* 2020 article, "This Is Inequity at the Boiling Point."[12] Similar to the "snapshots" in *Dry*, this article provides short features of places around the world that are experiencing extreme heat. Much like the shade privilege article, this article points out that "a hotter planet does not hurt equally. If you're poor and marginalized, you're likely to be much more vulnerable to extreme heat."[13] Yet again, students are faced with the reality of climate change inequities, and some even go back to their anchor text to consider how inequities are portrayed in *Dry*. They point out the families who are forced out of their homes as well as the families who have to migrate to other areas for relief. Students could read NPR's 2021 feature, "Americans Are Moving to Escape Climate Impacts. Towns Expect More to Come," to further explore the reality of climate migration both locally and internationally.[14] In *Dry*, many community members are forced to flee California and migrate to areas of the country with more water. Finally, the "Our Climate, Our Future" short climate story videos offer students a mentor text for "snapshots" they can present for or about their communities.[15] These videos feature young people sharing information about the specific ways in which climate change affects their communities. When students are given articles and other nonfiction formats to explore, they begin to localize the climate crisis within their communities.

YA Climate Fiction Text Study #2: *The Marrow Thieves* by Cherie Dimaline

> We are literally the land, a planet. Our spirits inhabit this place. We are not the only ones. We are creators of this place with each other. We mark our existence with our creations. It is poetry that holds the songs of becoming, of change, of dreaming, and it is poetry we turn to when we travel those places of transformation, like birth, coming of ages, marriage, accomplishments, and death.[16]

Studying a cli-fi novel that centers Indigenous voices of North America makes it abundantly clear that place matters. In the introduction of *When the Light Was Subdued, Our Songs Came through: A Norton Anthology of Native Nations Poetry*, Joy Harjo conveys the Native value of collectively creating place through stories. Through a study of Cherie Dimaline's *The Marrow Thieves*, students learn that just like all other events and experiences, a person's identity and relationship with place alters how they interact with the climate crisis. This is apparent as Frenchie and other Indigenous characters navigate their role in the climate crisis, and as they experience the fear of being hunted for their bone marrow due to the desperate actions of those who intensified the crisis. Dimaline portrays the "songs of becoming, of change, of dreaming" through the poetic weaving of "Story" throughout her novel. The importance of community, family, and storytelling arise as characters on the run are chased by those who have migrated due to rising sea levels and ensuing environmental disasters.

Reading The Marrow Thieves to Make Sense of the Climate Crisis

Our students' first read of *The Marrow Thieves* is supported by a discussion guided by the lens of race, place, and privilege. As we work with discussion leaders for weekly group meetings, we help them pose questions that center on identity and storytelling, specifically the role of storytelling among the Indigenous People of North America represented in the novel. Facilitators ask questions like "What is the purpose of dreams?," "What is the importance of community and what defines a community?," "How do people find hope in the midst of disaster?" Almost all of the student's questions and responses focus on collective action toward survival. The reality of climate change is not questionable or up for interpretation—Dimaline's world is already destroyed by it. Place and the natural world certainly serve as characters throughout the novel; however, much of the story focuses on storytelling itself. The characters rely on family, community, elders, and stories to survive. As students think about identity and place, they begin to discuss, ask questions, and research Indigenous storytelling, land, and traditions.

Reading The Marrow Thieves to Explore Climate Crisis "Form"

In an interview for Canadian TVO with Nam Kiwanuka, Dimaline addresses how her identity as a Métis writer affects her work. She notes that because

> [Indigenous People] are the people of stories, because we carry oral tradition, because it's such a high honor in the community to carry those stories, it is absolutely everything about how I write. The language I use, the structure of the story.[17]

Dimaline goes on to explain that even the way she introduces characters with "coming-to" stories provides the reader with an Indigenous "form" or way of telling stories. Her community is her identity which then informs the choices she makes when helping a reader make sense of a community and world unlike or similar to their own. We encourage students to think of how their own identities inform the way in which they tell stories and connect with their audience. In particular, we want students to think about how they fit into the collective identity of their community as they complete their community action projects.

Throughout our study of climate fiction, we consistently revisit the question "What role does storytelling have in creating positive change within communities?" Students often question if some "forms" of storytelling have a more obvious and immediate impact than other forms. Dimaline's intricate weaving of storytelling through every thematic idea in her novel offers students a clear example of an effective climate fiction technique. The use of storytelling to learn from elders and the past, to learn language, and to understand community, portrays an environmental disaster in which only those with dreams and communities to pass along stories survive. Dreams become a metaphor for hope.

In addition to weaving stories to communicate the pain of the past and the collective effort of survival in crisis, Dimaline uses nature imagery that gives readers access to a view of urban decay that may not be a familiar scene for all. As Frenchie and his family approach the Four Winds Resort, he reminisces that he "spent most of [his] early life in the crumbling end of the city, surrounded by urban decay and concrete waste where the skyline looked like a ruined mouth of rotted teeth."[18] Again, comparisons provide a better understanding of environmental disaster. Frenchie then describes entering the Four Winds Resort as "the moon lit the wide hall in pale ribbons, turning the dust and broken bits of chair and wainscotting and climbing vines from feral houseplants into fairy tale turrets."[19] Nature helps Dimaline describe the city that once was, and takes over the very space that the city inhabited.[20] Some students connected this scene to the first time they returned to previously closed schools or opened their school bags after the initial pandemic school closings as a similar feeling to Frenchie's description of entering a place of the past. Frenchie describes this feeling as "opening the lid of a sealed jar, and all the anticipation of a tomorrow planned a thousand yesterdays ago came skittering to our feet like slick-shelled beetles."[21] Once again, students notice the use of the comparison to help them make sense of the unknown as well as communicate a feeling of loss they already know well.

Pairing Nonfiction Texts with The Marrow Thieves

One of the risks we were aware of as we started this climate fiction unit was the possibility of our students experiencing "catastrophe fatigue" or "climate grief." Thus, we chose paired texts that focused on Indigenous stories and hope in the midst of catastrophe. In *The Atlantic* editorial "Inevitable Planetary Doom Has Been Exaggerated: Hope for the Future Is a Reasonable and Necessary Prerequisite for Action," Emma Marris acknowledges the feeling of doom that stems from environmental writing and climate change education. However, she also emphasizes the value of local action and the prioritization of local policy. Our students know and often care deeply about their personal communities. They are also able to see immediate impacts of local policy. In fact, our literary focus on race, place, and privilege offers a lens of hope. In her editorial, Marris confirms that "many of our problems are so thoroughly tangled up with one another that we may not need to fight them separately. Environmental destruction disproportionately harms people of color and lower-income people." Thus,

> fighting for racial or economic justice, or against voter suppression, still can mean fighting for the environment. As these links are becoming better understood, the environmental movement is finally working with its natural allies to, for example, fight fossil fuels while promoting investment in Black, Indigenous, brown, and working-class communities.[22]

Through a focus on converging crises, students realize that any action toward justice can create progress, especially at the local community level. This realization helps students see that the topics chosen in their community projects are not unrelated or less important than climate change. In fact, their work toward justice or change in one area is connected and important to their understanding of the climate crisis.

In addition to articles about hope, we chose articles that center Indigenous voices. We want students to consider whose voices are a part of climate solution conversations and whose voices are left out. Additionally, we challenge students to identify which groups of people are predominantly affected by the negative effects of climate change. *The New York Times* article "Indigenous People Are Changing Sci-Fi" by Alexandra Alter offers another feature of the value Indigenous storytellers bring to writing about the environment. They often present a more earth-centered perspective that impacts readers in a different way as they try to understand literary worlds changed by climate crises. The article features several current Indigenous writers, including Rebecca Roanhorse. According to Roanhorse, she uses science fiction and apocalyptic narratives to "depict a world where Native culture, language and people have endured, in spite of efforts over the centuries to wipe them out."[23] The article also features Cherie Dimaline who states, "The only way I know who I am and who my community is, and the ways in which we survive and adapt, is through stories."[24] Climate stories offer models of survival and hope in the face of converging crises.

The Alliance for Climate Education (ACE) *Youth Think Climate* (YTC) magazine includes stories of climate activism and education within communities of young people. Their February 2021 issue highlights the voices and people of climate justice initiatives. Their interview with Katherine Quaid, "Nez Perce, Cayuse, a Citizen of the Confederate Tribes of Umatilla," offers students a glimpse into the life and thoughts of a climate activist. As the Communication and Outreach Coordinator for the Women's Earth and Climate Action Network (WECAN), Quaid states that she fights for climate justice in order to "fight for [her] communities, [her] loved ones, and for the land of [her] ancestors, and all animal and plant beings." She also argues that "we must address the root causes of our climate crisis—Colonization, White Supremacy, Capitalism, Patriarchy—if we genuinely want to build the world we seek."[25] Through the *YTC* magazine interviews, students see more examples of young people working toward an investigation of root cause, an act we expect from them as they complete their community action projects.

YA Climate Fiction Text Study #3: *Not a Drop to Drink* by Mindy McGinnis

Mindy McGinnis's YA climate fiction starts with a nine-year-old girl killing a man. The opening scene casts a dire immediacy of the adultification of (particularly) young girls in imagined dystopian futures. Lynn, the book's protagonist, continues through the text with the nervous feminine energy of a woman who loses the minimal protection of a structured society. Lynn's isolation and the ever-present threat of violence reflect the aspect of the climate crisis that often goes overlooked: the idea that the most vulnerable populations will be affected by the most drastic of consequences in an environmental collapse.

Not a Drop to Drink is a novel that follows Lynn—a young girl living in not-so-distant future Ohio—along a journey to protect the pond that has supplied her with potable water since she remembers. When her mother dies, she has to forge human connection, an action that has always been the most dangerous in a space where people kill on sight in order to either procure or protect dwindling water sources. Her struggle reflects the evolutionary psychology that tries to anticipate how humans will act in survival situations. The place that she inhabits both sustains her life and puts her at risk of danger as her privileged proximity to water makes her a prime target for many others suffering without her positional advantage. The novel portrays Lynn as a character of juxtapositions: at one moment privileged, living in a protective bubble away from the world that has disintegrated from environmental degradation and ensuing chaos, and in another as naive and secluded from the realities of the outside world. Throughout an exploration of the novel and

Lynn's character, the ideas of place and privilege evolve into driving forces of plot and cultural examinations of speculative futures.

Combined with nonfiction and multimodal texts, Mindy McGinnis's *Not a Drop to Drink* allows readers to make sense of rural poverty versus governmentally controlled cities, the psychology of scarcity, the insecurity of female isolation, and the processes of human connection that either evolve or devolve in the face of catastrophe. Although a speculative piece, McGinnis's novel acts as a handhold for understanding the uncertainty produced by climate change even today.

Reading Not a Drop to Drink to Make Sense of the Climate Crisis

Lynn and her neighbors in *Not a Drop to Drink* are positioned as lower socioeconomic status in that they have chosen to live independently, off the grid of the cities that function under keen governmental eyes with corrupt water rationing practices. Throughout the text there are insinuations of the simplicity or rugged quality of Lynn and the people who chose to live outside the established society of her dystopian world. In this way, Lynn's character reflects the reality of the climate crisis as having a disproportionate effect on populations with less access to resources, despite having the least culpability. A recent Oxfam report sums up the convergence that Lynn experiences:

> While no one is immune [to climate change], the consequences will continue to be more devastating for low and middle-income countries, which are most at risk and face the highest climate-fueled displacements, and within these countries for women living in poverty, who typically bear greater responsibility for tasks that are made more difficult by climate change, including sourcing food and water.[26]

Although *Not a Drop to Drink* is set in the United States, there is a vast distinction between more rural, less developed portions of America and the industrialized national dream. While students can read a report that states the truths, engaging with the complicated daily process of Lynn's water purification system creates a much more compelling story:

> It would be a purifying day for sure, which meant hours of labor. She pushed the lip of her first bucket under the surface of the water, trying not to disturb the muddy bottom. No matter how careful she was, there were always flecks of dirt and algae that settled in the holding tanks. She moved along the bank to a new spot to dip the second bucket. . . . She twisted the plastic cap off the top of the tank and dumped both buckets into it, listening to the tone of the falling water change as the level rose. This tank was the unpurified pond water. The

> other stood half full of water that had already been rotated out to the tin sheets, and would be drawn off through the winter to fill the small thousand-gallon tank that was in the basement.[27]

The simple statement of excessive climate-induced labor beleaguering women and low-income families more heavily becomes an all-day, repetitive struggle of survival when told as a story rather than a fact. Young women can imagine themselves toiling over a task that in current times is as simple as turning a handle on a faucet much more intensely when it combines both scientific reasoning and narrative emotion. This is the power of storytelling and world-making that allows students to become immersed in a possible future, as opposed to dismissing something that feels distant and disconnected.

Not a Drop to Drink and other climate fiction texts can also offer persuasive evidence toward solution-oriented research and suggestions that come from nonfiction sources. In McGinnis's text there are allusions to resource hoarding and government control as Lynn meets people from the walled-in city. The young character Lucy states that "they don't let everybody see those [water] maps though, even in the city. Only soldiers get to look, and even then only the super special ones."[28] Young students can understand a shady, villainous force keeping secrets from the populace as an entertainment trope more readily than the actual truth of their lived reality where "in country after country it is the richest who are least affected by the pandemic, and are the quickest to see their fortunes recover. They also remain the greatest emitters of carbon, and the greatest drivers of climate breakdown."[29] While our students are aware of power differentials and inequity in their governing bodies, there is still an innate trust that those with control have the best interests of the masses in mind. Climate fiction makes subversive truth more palatable and less shattering for young adults, introducing them to ideas of privilege without destroying optimism. The implication of corruption in a dystopian novel allows young people to imagine (and thus problem-solve) stories in their own realities.

In another attempt to convince young people to adhere to the scientific reasoning offered today, climate fiction portrays imagined futures where individuals have ignored proposed practices. Lynn's experience is dictated by water scarcity and it remains the environmental consequence with the most impact in her story. In their local areas, students can study similar drought-ridden areas and proposed solutions in order to craft a future different from Lynn's tumultuous existence. For example, "nearly 40 million people rely on the Colorado River system, which is experiencing a historic 16-year dry period with models suggesting that flows could fall by as much as 55 percent by 2100."[30] While this statement may be obscure to the developmental stage of a middle schooler's brain, in conjunction with passages about the stream settlements of Lynn's neighbors, the harsh reality of water scarcity, and its connection to place, its implications become evident. Lynn meets a family settled by the stream near her pond living in dire nomadic circumstances; they have traveled from their place in the city to reside by a water source, ignoring the danger of the cold and the openness to attack. Like those living now in the heat belt or in places fed by rivers, the travelers must make choices about life quality based on their geographic position and societal structure, a reflection of the real lived consequences of the climate crisis.

Students can use climate fiction to either mirror the climate crisis reality or to interrupt it with future-forming based on the fusion of scientific warnings and proposed effects. Through a paired reading of *Not a Drop to Drink* with nonfiction texts students can process the effects of the climate crisis in a way that feels safe and unobtrusive, ushering them into a deeper understanding of the issue without a feeling of overwhelming dread. This sensemaking process can lead to discussion-based literary activities tied into research practices. With a combination of climate fiction and climate science, students are empowered to consume and create stories of tragedy but also resilience. Lynn's story centers the young female perspective with a focus on her vulnerability, responsibility, and the burden of world recreation, offering a window for students to view a world that could possibly be.

Reading Not a Drop to Drink *to Explore Climate Crisis "Form"*

In climate fiction, and the wider dystopian genre in general, scenes with descriptive action and intense imagery build vivid stories for students that make concrete the immediate threat of the climate crisis. McGinnis's writing is no exception as it offers visions of a bleak future reality while appealing to readers' empathy for characters with whom they can relate. Throughout the text students can note, analyze, and model their own writing after intense scenes that reflect dramas of the climate crisis that many students do not have first-hand experience with in their own lives.

In our own unit, we asked students to emulate two different evocative scenes, one with a focus on striking imagery and one for practice with the descriptive power of verbs and dialogue. Asking students to use data and truth to create scenes of particular climate crisis consequences is a technical exercise in drawing inspiration and processing different modes of information, combining storytelling and sensemaking skills.

One particular example of vivid imagery that personalizes the climate crisis is the scene in which the young character Lucy has to have her shoes removed after not being able to take them off while walking the long distance from the city center, to Lynn's neighborhood, and staying by the damp stream. McGinnis writes,

> Her feet were a mess. Dead skin hung in flaps from blisters long since burst, a fungal infection covered most of her left sole, and all her toenails had grown inward in response to the shoes that hadn't left her feet. It was a miracle the girl would still walk.[31]

Using this as a model text, students were asked to choose for their own writing a particularly graphic, yet more minor, example of the consequences of climate change that are not reported on but would surely affect individuals. Whether grotesque or tragic, scenes like this serve a dual purpose again of analysis (asking students to imagine how even the particulars of appearance and comfort would be affected by climate crisis) and generation (asking students to write their own descriptions of small instances that would make up their lived reality in a climate-changed world). In this exercise, students not only utilized their empathy to analyze examples of the climate crisis effects but also conceptualized possibilities of the crisis through their own fiction workshopping.

The final scenes of *Not a Drop to Drink* exemplify the chaos of the climate crisis by creating a microcosm of violence incited by climate consequences. Lynn encounters the dangers of water scarcity and resource hoarding daily, but the resolution of the story is a raging outburst that students can analyze as metaphorical or literal. In the scene, Lynn is waging an attack on a corrupt and misogynistic settlement that she is attempting to disassemble. During the attack,

> Stebbs and Lynn fired at the same time, her crosshairs trained on Father. He fell, clutching a shattered shoulder. His hand dangled lifeless from the dead arm, his gun useless on the group. The hall guard dropped to his knees and fired at Eli before Stebbs' bullet could reach him. The guard's brain exploded through the back of his head, but not before his bullet hit Eli's backpack . . . Eli became a living ball of fire.[32]

Students can analyze the fast-paced, action-packed violence of this scene along with its urgent tone as a measure of the devolution of societal safety that could result from the dissolution of a society facing climate crisis. Those consequences do not have to be weather or environmentally applicable to reach into the very fabric of social mores and behaviors. Such scenes highlight for students the vast lifestyle changing repercussions our hesitation on climate issues may have, allowing students to trace root causes to branches of impact.

Pairing Nonfiction Texts with Not a Drop to Drink

When reading *Not a Drop to Drink*, there are a few ancillary texts that can aid both student engagement as well as connect students to the urgency of the climate crisis in real life. One example of these texts is Ibram X. Kendi's piece in *The Atlantic* entitled "What the Believers are Denying." This thought piece explores the connection between denial of scientific fact in both the climate crisis and systemic racism; it demonstrates that while both issues have observable evidentiary support many still refuse to believe in or act upon them. Kendi draws conclusions around the ways in which people are presented with proof of phenomena but continue to act out of alignment with the truth. This piece asks students to switch their thinking and suggests that "instead of asking, 'Are you a racist?' we should be asking, 'What is a racist?' Instead of asking, 'Do you believe in climate change?' we should be asking, 'What does climate change look like?'" to take back power from deniers.[33]

Another piece that presents students with the challenge of forward thinking is from *Slate* by Duke Reiter and Wyatt Scott titled "The Heat Belt Is a Glimpse into a Climate-Changed America." This article allows students to bridge the connection between fiction and nonfiction, current reality and dystopian prediction, by extrapolating present data from areas in the United States most affected by climate change to show possible future conditions. The piece also encourages students to be active participants in their communities' reactions and preventive climate strategies as it opines that the most effective measures are community-driven.

Although students will benefit from recently updated and newer pieces of text during their study of YA climate fiction, another helpful text is CNBC's examination of an Oxfam report titled "The '1%' Are the Main Drivers of Climate Change, but It Hits the Poor the Hardest: Oxfam Report." This nonfiction analysis presents an economic point of view on climate change, focusing on the privileged wealth of the predominant producers of the climate crisis and the climate consequences experienced by those who have less. This piece also connects the climate crisis to other privilege-influenced natural disasters like global pandemics which encourages students to apply the framework of race, place, and privilege to many instances of distress outside of the climate environment.

COMMUNITY ACTION

The YA climate fiction genre encourages action from young people. After reading for both understanding and form, we ask students to revise their community projects. We guide them through a consideration of the intersecting problems that are at play within their communities. Students work individually or within small groups to create and enact a plan for awareness and change. Students then present their findings and reflections to their peers, families, and local leaders. Although not all students will choose climate-focused issues in their community, they now have a clear understanding of the ecological nuances that exist within it. Through cli-fi characters students learn that there is not one cause of any singular societal issue, but that they are, in fact, interconnected, often a product of underlying interrelated systemic problems. Before students determine their project audience, form, or specific product, we ask each group to complete a root cause analysis of their community problem. We encourage them to apply the systems lens that they used in their readings to their work in communities. In her book, *Thinking in Systems*, Donella Meadows states that, "as our world continues to change rapidly and become more complex, systems thinking will help us manage, adapt, and see the wide range of choices we have before us."[34] As we help students understand through the literature, "systems happen all at once. They are connected not just in one direction, but in many directions simultaneously."[35] Throughout the unit, we explored these concurrent systems and the ways in which writers take risks with their form and techniques in order to represent science, thus allowing the reader to make sense of their world.

With their problem, root cause, and audience in mind, we ask students to take risks as they encourage change by taking action in their communities. Some products included bringing awareness to fast fashion through a thrift store fashion show; podcasts about justice issues in the community with an episode focusing on climate justice; op-eds, poems, photography galleries in school halls to emphasize humans' impact on the environment; restorative circles with local police officers and community members; and QR codes with climate change facts posted around the school campus. The possibilities for community involvement through environmental nonprofits and local leaders also allow students to begin to see their role in a more sustainable and justice-centered community. At the local level, students can critically think about the role of race, place, and privilege in their own lives and in their surrounding communities. They can experiment with the most effective storytelling forms that portray climate justice. Through this risk-taking in their writing and activism, students can discover what collective environmental action looks like in their community. In time, revision in their writing and their products becomes a tool to create lasting change.

CONCLUSION

Through student choice of high-interest climate fiction and nonfiction pairings, students see the intersectionality of the climate crises, young people as agents of change, and effective local community efforts. In *Between the World and Me*, Ta-Nehisi Coates makes these interconnecting crises clear:

> Once, the Dream's parameters were caged by technology and by the limits of horsepower and wind. But the Dreamers have improved themselves, and the damming of seas for voltage, the extraction of coal, the transmuting of oil into food, have enabled an expansion in plunder with no known precedent. And this revolution has freed the Dreamers to plunder not just the bodies of humans but the body of the Earth itself.[36]

Rather than politicizing or questioning climate change, YA climate fiction gives students an opportunity to understand science through stories that value the power of young people. As English Language Arts educators, we hope that these stories will help students understand that the climate crisis stems and is compounded by what Ta-Nehisi Coates would say are "phenomena [that] are known to each other":

> It was the cotton that passed through our chained hands that inaugurated this age. It is the flight from us that sent them sprawling into the subdivided woods. And the methods of transport through these new subdivisions, across the sprawl, is the automobile, the noose around the neck of the earth, and ultimately, the Dreamers themselves.[37]

We want stories to illuminate the confluence of societal and systemic issues that lead to necessary climate justice conversations, initiatives, and action.

At the time we are writing this chapter, we are in the midst of a global pandemic and newly emerging climate crises where connections to climate fiction are eerily familiar. However, through our intentional study of storytelling, climate grief and catastrophe fatigue are countered with imagined literary scenarios that offer hope and action for students. In our current moment, it is difficult for students to disconnect the value of community action in their climate fiction studies with the communal efforts needed to make it through present crises. Ultimately, we want to offer students opportunities to process the current and future unknowns through the power of storytelling. We want students to recognize that storytelling is the process of showing others what it means to be human among disaster.

NOTES

1. Ibram X. Kendi, "What the Believers Are Denying," The Atlantic, January 1, 2019, https://www.theatlantic.com/ideas/archive/2019/01/what-deniers-climate-change-and-racism-share/579190/.

2. Dorothy Roberts, *Fatal Invention: How Science, Politics, and Big Business Re-create Race in the Twenty-First Century* (New York: The New Press, 2012), 2.

3. Ijeoma Oluo, *So You Want to Talk about Race* (New York: Hachette Book Group, 2019), 59.

4. Kendi, "Believers."

5. "Resolution on Literacy Teaching on Climate Change," *NCTE*, March 1, 2019, https://ncte.org/statement/resolution-literacy-teaching-climate-change/ - :~:text=Teaching climate change to adolescents, at the 2018 Annual Convention.

6. Richard Beach, Jeff Share, and Allen Webb, *Teaching Climate Change to Adolescents: Reading, Writing, and Making a Difference* (New York and Urbana, IL: Routledge and NCTE, 2017), 7.

7. Anna Bernstein and Kaela Sweeney. "The Climate Crisis Confluence: How Race, Place, and Privilege Flow Together in YA Literature" (presentation, National Council of Teachers of English Virtual Annual Convention, November 19-22, 2020), https://vimeo.com/523470012.

8. Jarrod Shusterman and Neal Shusterman, *Dry* (New York: Simon & Schuster, 2018), 3.

9. Shusterman and Shusterman, *Dry*, 76.

10. Shusterman and Shusterman, *Dry*, 76.

11. Tim Arango, "'Turn Off the Sunshine': Why Shade Is a Mark of Privilege in Los Angeles," *New York Times*, December 1, 2019, https://www.nytimes.com/2019/12/01/us/los-angeles-shade-climate-change.html.

12. Somini Sengupta, "This Is Inequity at the Boiling Point," *New York Times*, August 7, 2020, https://www.nytimes.com/interactive/2020/08/06/climate/climate-change-inequality-heat.html.

13. Sengupta, "Boiling Point."

14. Annie Ropeik, "Americans Are Moving To Escape Climate Impacts. Towns Expect More To Come," *NPR*, January 22, 2021, https://www.npr.org/2021/01/22/956904171/americans-are-moving-to-escape-climate-impacts-towns-expect-more-to-come.

15. "Climate Stories," Our Climate, Our Future, accessed February 10, 2021, https://ourclimateourfuture.org/map/.

16. Harjo, Joy, ed., *When the Light of the World Was Subdued, Our Songs Came Through: A Norton Anthology of Nations Poetry* (New York: W.W. Norton & Company), 1.

17. Cherie Dimaline, "Reclaiming Lost Dreams." Interview by Nam Kiwanuka, The Agenda with Steve Paikin, TVO, August 22, 2018, audio, 20:10, https://www.tvo.org/video/reclaiming-lost-dreams.

18. Cherie Dimaline, *The Marrow Thieves* (Altona: Cormorant Books Inc., 2017), 59.

19. Dimaline, *The Marrow Thieves*, 60.

20. See also Alan Weisman's *The World Without Us*.

21. Dimaline, *The Marrow Thieves*, 60.

22. Emma Marris, "Inevitable Planetary Doom Has Been Exaggerated," *The Atlantic*, February 1, 2021, https://www.theatlantic.com/science/archive/2021/02/other-side-catastrophe/617865/s.

23. Alexandra Alter, "'We've Already Survived an Apocalypse': Indigenous Writers Are Changing Sci-Fi," *New York Times*, August 14, 2020, https://www.nytimes.com/2020/08/14/books/indigenous-native-american-sci-fi-horror.html?smid=tw-share.

24. Alter, "Apocalypse."

25. Katherine Quaid, "An Interview with Katherine Quaid," *Youth Think Climate*, Issue 2: *Climate Justice*, February 10, 2021, https://acespace.org/2021/02/10/youth-think-climate-issue-2/.

26. Catherine Clifford, "The '1%' Are the Main Drivers of Climate Change, But It Hits the Poor the Hardest: Oxfam Report," *CNBC*, January 27, 2021, https://www.cnbc.com/2021/01/26/oxfam-report-the-global-wealthy-are-main-drivers-of-climate-change.html.

27. Mindy McGinnis, *Not a Drop to Drink* (New York: Katherine Tegen Books, 2013), 6–7.

28. McGinnis, *Not a Drop to Drink*, 151.

29. Clifford, "Oxfam."

30. Duke Reiter and Wyatt Scott, "The Heat Belt Is a Glimpse Into a Climate-Changed America," *Slate*, February 18, 2021, https://slate.com/technology/2021/02/heat-belt-climate-change-local-action-interviews.html.

31. McGinnis, *Not a Drop to Drink*, 94.

32. McGinnis, *Not a Drop to Drink*, 295.

33. Kendi, "Believers."

34. Donella H. Meadows, *Thinking in Systems* (White River Junction: Chelsea Green Publishing, 2008), 2.

35. Meadows, *Thinking in Systems*, 5.

36. Ta-Nehisi Coates, *Between the World and Me* (New York: Spiegel & Grau, 2015), 151.

37. Coates, *Between the World and Me*, 151.

REFERENCES

Alter, Alexandra. "'We've Already Survived an Apocalypse': Indigenous Writers Are Changing Sci-Fi." *New York Times*, August 14, 2020. https://www.nytimes.com/2020/08/14/books/indigenous-native-american-sci-fi-horror.html?smid=tw-share.

Arango, Tim. "'Turn Off the Sunshine': Why Shade Is a Mark of Privilege in Los Angeles." *New York Times*, December 1, 2019. https://www.nytimes.com/2019/12/01/us/los-angeles-shade-climate-change.html.

Beach, Richard, Jeff Share, and Allen Webb. *Teaching Climate Change to Adolescents: Reading, Writing, and Making a Difference*. New York and Urbana, IL: Routledge and NCTE, 2017.

Bernstein, Anna, and Kaela Sweeney. "The Climate Crisis Confluence: How Race, Place, and Privilege Flow Together in YA Literature." Conference Presentation at the National Council of Teachers of English Virtual Annual Convention, November 2020. https://vimeo.com/523470012.

Clifford, Catherine. "The '1%' Are the Main Drivers of Climate Change, But It Hits the Poor the Hardest: Oxfam Report." *CNBC,* January 27, 2021. https://www.cnbc.com/2021/01/26/oxfam-report-the-global-wealthy-are-main-drivers-of-climate-change.html.

Coates, Ta-Nehisi. *Between the World and Me.* New York: Spiegel & Grau, 2015.

Dimaline, Cherie. *The Marrow Thieves.* Altona: Cormorant Books Inc., 2017.

Dimaline, Cherie. "Reclaiming Lost Dreams." Interview by Nam Kiwanuka. *The Agenda with Steve Paikin*, TVO, August 22, 2018. Audio, 20:10. https://www.tvo.org/video/reclaiming-lost-dreams.

Harjo, Joy, ed. *When the Light of the World Was Subdued, Our Songs Came Through: A Norton Anthology of Nations Poetry.* New York: W.W. Norton & Company, 2020.

Kendi, Ibram X. "What the Believers Are Denying." *The Atlantic,* January 1, 2019. https://www.theatlantic.com/ideas/archive/2019/01/what-deniers-climate-change-and-racism-share/579190/.

Marris, Emma. "Inevitable Planetary Doom Has Been Exaggerated." *The Atlantic*, February 1, 2021. https://www.theatlantic.com/science/archive/2021/02/other-side-catastrophe/617865/.

McGinnis, Mindy. *Not a Drop to Drink.* New York: Katherine Tegen Books, 2013.

Meadows, Donella H. *Thinking in Systems.* White River Junction: Chelsea Green Publishing, 2008.

NCTE. "Resolution on Literacy Teaching on Climate Change." March 1, 2019. https://ncte.org/statement/resolution-literacy-teaching-climate-change/#:~:text=Teaching%20climate%20change%20to%20adolescents,at%20the%202018%20Annual%20Convention.

Oluo, Ijeoma. *So You Want to Talk about Race.* New York: Hachette Book Group, 2019.

Our Climate, Our Future. "Climate Stories." Accessed February 10, 2021. https://ourclimateourfuture.org/map/.

Quaid, Katherine. "An Interview with Katherine Quaid." *Youth Think Climate, Issue 2: Climate Justice*, February 10, 2021. https://acespace.org/2021/02/10/youth-think-climate-issue-2/.

Reiter, Duke and Wyatt Scott. "The Heat Belt Is a Glimpse Into a Climate-Changed America." *Slate,* February 18, 2021. https://slate.com/technology/2021/02/heat-belt-climate-change-local-action-interviews.html.

Ropeik, Annie. "Americans Are Moving to Escape Climate Impacts. Towns Expect More to Come." *NPR*, January 22, 2021. https://www.npr.org/2021/01/22/956904171/americans-are-moving-to-escape-climate-impacts-towns-expect-more-to-come.

Rothenberg, Paula S. *White Privilege: Essential Readings on the Other Side of Racism.* New York: Worth Publishers, 2005.

Sengupta, Somini. "This Is Inequity at the Boiling Point." *New York Times*, August 7, 2020. https://www.nytimes.com/interactive/2020/08/06/climate/climate-change-inequality-heat.html.

Shusterman, Jarrod, and Neal Shusterman. *Dry.* New York: Simon & Schuster, 2018.

Chapter 4

Starting Points for Student Inquiry into Our Relationship with the Environment

Ryan Skardal

British Columbia's English Studies 12 curriculum, which guides my teaching, reminds teachers that "the exploration of text and story deepens our understanding of diverse, complex ideas about identity, others, and the world."[1] This standard admirably asks students to consider their identity within the world, and it therefore urges teachers of English to develop their awareness and implementation of ecocriticism. In his book, *Ecocriticism*, Greg Garrard defines the titular critical framework as "the study of the relationship of the human and the non-human, throughout our cultural history and entailing critical analysis of the term 'human' itself."[2] It is good that BC's Ministry of Education urges students to explore their identity and their understanding of the world because the climate crisis suggests that our current models for understanding who we are within the physical environment are flawed, at best.

In considering ways to implement ecocritical ideas into my practice, I first turned to my own time as a student and initially did not find much on which to build—until I realized that we have been teaching students about their relationship with the environment for decades. When I was a student, for example, my teachers taught us "person versus nature" units. We read stories like Gary Paulsen's *Hatchet* or Jack London's "To Build a Fire." In these texts, gritty heroes survive against all odds in the face of brutal and unforgiving wilderness. They learn to bend the wild world to their will or else they perish. Should these heroes return to "civilization," their relationship with the physical environment is altered, becoming either an afterthought or a fantasy of adventure. The physical environment can indeed be dangerous and a source of adventure for people, as readers see when they read memoirs like Jon Krakauer's *Into Thin Air* or Liz Clark's *Swell*, though sometimes it is also depicted as beautiful and inspiring. These units of study were moving, but the conclusions most easily reached in them did little to address the climate crisis. We should reimagine these "person versus nature" units in the face of the climate crisis.

It is difficult to fully describe the complexity and enormity of the climate crisis. Timothy Morton identifies global warming as a "hyperobject" because "such things are impossible to point to all at once."[3] The effects of climate change are everywhere, and yet so massive that we struggle to understand them fully or meaningfully. Because of our choices, especially those made by the developed world, atmospheric carbon dioxide concentrations and ocean acidification are rising while biodiversity levels are falling. Oceans are rising, glacial ice is melting, and too often forests are either burning or being cut down unsustainably. Every day, we can find stories about environmental destruction or disaster in the news, and yet it is difficult to see a deep awareness and responsibility for the choices that have led us here and that exacerbate the climate crisis. It is increasingly common to hear people say that we live in the "Anthropocene," a word popularized by Paul Crutzen, which describes an era in which the surface of the earth is dominated by human choices. There is a clear need to respond to the climate crisis, so how should these studies start when addressing such a broad and wide-ranging topic?

Because teachers of English routinely explore identity, they might find it useful to start with this concept. It is fitting, and perhaps ironic, that the Anthropocene concept describes a "world dominated by human choices" because that phrase also captures the implicit ethos of the global economy's dominant environmental ethos, anthropocentric cornucopianism.[4] In the cornucopian model, the world's resources are abundant—treated as limitless—and they are for humanity to do with as they wish. We see the special status of people, with their conscious minds, in the brainy background of Michelangelo's *The Creation of Adam* as well as in Stewart Brand's famous declaration in *The Whole Earth Catalogue* that "we are as gods and might as well get used to it."[5] In *The Cosmopolitan Tradition: A Noble but Flawed Ideal*, Martha C. Nussbaum argues that this anthropocentric ethos assigns value to the well-being of all people, but it excludes nonhuman animals and the planet from consideration—a dualist model that prioritizes mind over matter. Nussbaum traces the philosophical assumptions of this ethos to Diogenes and a "Cynic/Stoic cosmopolitanism [that] urges us to recognize the equal, and unconditional, worth of all human beings, a worth grounded in moral choice-capacity [. . .], rather than on traits that depend on fortuitous natural or social arrangements."[6] The cosmopolitan tradition is admirable in its potential capacity to assign human dignity to all people, rather than just men or the rich or members of one's in-group. Nussbaum sees in its conception of human dignity "one endpoint of a line that leads to the modern human rights movement,"[7] so it would be difficult to complain without qualification about such a tradition.

However, because we concern ourselves only with the well-being of people, we are left with a framework in which ecological loss does not intrinsically matter. Instead, only the physical world's utility, or its ability to serve or harm people, matters. At first glance, the trade-off has been great for many. In *Utopia for Realists*, Rutger Bregman points out that

> billions of us are suddenly rich, well nourished, clean, safe, smart, healthy, and occasionally even beautiful. Where 84% of the world's population still lived in extreme poverty in 1820, by 1981 that percentage had dropped to 44%, and now, just a few decades later, it is under 10%. If this trend holds, the extreme poverty that has been an abiding feature of life will soon be eradicated for good.[8]

Bregman outlines a medieval utopia, Cockaigne, or the Land of Plenty, and argues that much of the world has realized that dream. Through a variety of programs, including the Sustainable Development Goals, people around the world work to help underdeveloped economies address poverty and develop wealth, but what about the costs of climate change in the coming years and decades? In *Strategies of Commitment*, Thomas Schelling points out that undeveloped economies are more likely to rely on agriculture and therefore more likely to suffer the adverse effects of climate change on the economy; therefore, "the worst effects of deteriorating climate can be avoided if poor countries can become nonpoor in the coming half-century."[9] For Schelling, investing today in climate mitigation to benefit future generations, or perhaps even ourselves in a few years, is "very much like a foreign aid program with some of the foreigners being descendants who live not on another continent but in another century."[10] Following this train of thought, the economist William Nordhaus concludes that his economics and climate model (the DICE model) suggests that

> a 2 °C target is both too low and too high. It is too low given the identified damages analyzed [. . .] and the high costs of attaining such an objective. [. . .] But it is too high if we believe, along with many earth scientists, that the earth has already crossed the thresholds of some of the dangerous tipping points.[11]

This train of thought would have us weigh the tourist dollars generated by coral reef against the cost of mitigating greenhouse gases or the present-day burning of coal against the losses of a potentially wealthier economy. Whether such actions work for planetary systems is almost irrelevant under this framework.

We are now confronting the limitations and the ecological trade-offs we have made because of our anthropocentric model. Nussbaum points out that "its disdain for non-human animals and the world of nature" is a problem because the "tradition grounds our duties in the worth and dignity of moral/rational agency."[12] It is this special status accorded in how we conceive of ourselves that permits people to exploit the physical world for their own gain, and it is apt that Kimberly Nicholas says we have an "exploitation mindset."[13] How could we not? I look at progress being made in the Sustainable Development Goals or at Hans Rosling's facts about falling infant mortality and feel grateful. And yet there is something absurd to me about an ethos that finds if the costs imposed on the economy by climate change are less than economic growth, then climate change is "worth it." Regardless, that is the system we have made.

Heterodox economists like Kate Raworth are attempting to devise new economic models, but English teachers will be more likely to employ ecocritical frameworks that "explore and even resolve ecological problems in [a] wider sense."[14] The ecocritical framework most familiar to a lay audience is almost certainly "environmentalism," which captures

> the very broad range of people who are concerned about environmental issues such as global warming and pollution, but who wish to maintain or improve their standard of living as conventionally defined, and who would not welcome radical social change. [. . .] Many value rural ways of life, hiking or camping, or are members of one of the mainstream environmental organisations.[15]

This group is easily criticized as anthropocentric cornucopianism with some measure of ecological conscience, which is why it is sometimes referred to as "shallow" environmentalism. Today, much of the political energy in the American response to climate change can be seen in the Green New Deal, which unites concerns from ecofeminists, eco-Marxists, and environmental justice values, but their actions broadly fit within the anthropocentric framework of shallow environmentalists. Garrard distinguishes environmentalism from deep ecology:

> Whereas "shallow" [environmentalist] approaches take an instrumental approach to nature, arguing for preservation of natural resources only for the sake of humans, deep ecology demands recognition of intrinsic value in nature. It identifies the dualistic separation of humans from nature promoted by Western philosophy and culture as the origin of environmental crisis, and demands a return to a monistic, primal identification of humans and the ecosphere. The shift from a human-centred to a nature-centred system of values is the core of the radicalism attributed to deep ecology, bringing it into opposition with almost the entirety of Western philosophy and religion.[16]

Many cultures revere the physical world, or aspects of it, even if our global economy continues to operate within an at best watered-down form of cornucopianism. To teach within an ecocritical framework is to nudge students toward potentially new ways of thinking about themselves within the physical world.

What realizations should teachers reasonably expect students to come to in the face of these changes? After Greta Thunberg became a high-profile climate activist, it seemed natural to expect my students to follow in her footsteps. In looking at climate science, Thunberg came to a new realization about humanity's role within the world, which she recounts in her book, *No One Is Too Small to Make a Difference*:

> They keep saying that climate change is an existential threat and the most important issue of all. And yet they just carry on like before. If the emissions have to stop, then we must stop the emissions. To me, that is black and white.[17]

My students mostly don't make such a drastic switch, and I now view it as an unrealistic expectation. Instead, I think more often now about the five stages of nonlinear "radical climate acceptance" Kimberly Nicholas describes in *Under the Sky We Make*: ignorance, avoidance, doom, all the feels, and purpose.[18] When I speak to my students, they often acknowledge that climate change is a "big deal" but will also express exhaustion about the topic, saying things like "I'm tired of hearing about this stuff." Even if they do consider climate change a climate "crisis," they are still very likely to "engage in a kind of climate change denial. [They] look for a split second and then [they] look away."[19] I don't view these statements from students as an indictment of my students so much as an indication that they have grown up in a developed economy founded on anthropocentricism and powered by fossil fuels. To live in this framework requires people to psychologically wall themselves off from the physical world and the way our choices affect it for the worse. Naomi Klein argues that "living with this kind of cognitive dissonance is simply part of being alive in this jarring moment in history, when a crisis we have been studiously ignoring is hitting us in the face."[20] It is frightening and shameful to consider the full implications of how we conceive of ourselves within the context of the physical world, so we try not to think about it.

Although potentially uncomfortable, teachers should still reframe "person versus nature" units to foster an interrogation into humanity's relationship with the physical world within the context of the climate crisis, and we can thankfully point to many models of inquiry in which thinkers consider how to change their personal ethos in response to greater environmental awareness. In *This Changes Everything*, Naomi Klein shares how she long accepted climate science without really changing her life at all. This thinking allowed her to psychologically separate herself from the physical world until she experienced a moment of awakening to the climate crisis. In her words,

> I found that I no longer feared immersing myself in the scientific reality of the climate threat. I stopped avoiding the articles and the scientific studies and read everything I could find. I also stopped outsourcing the problem to the environmentalists, stopped telling myself this was somebody else's issue, somebody else's job.[21]

In *The Omnivore's Dilemma*, Michael Pollan begins his interrogation into American food chains reminding himself that, "like every creature on earth, humans take part in a food chain, and our place in that food chain, or web, determines to a considerable degree what kind of creature we are."[22] In that book, Pollan explores the "food chains" that sustain him and ultimately strives to reconnect to the land by hunting, gardening, and mushroom gathering. In *Animal, Vegetable, Junk*, Mark Bittman argues that

> the old ways had been sacrificed to give birth to a new god—what's euphemistically called the market economy, or what we know as unrestrained capitalism. It would take Western science centuries to develop a truly rational branch of thinking, one that recognizes that everything is connected—the body, the natural and spiritual worlds, the wondrous and the inexplicable and the irrational. That branch of thought, the opposite of Cartesian dualism, is called ecology.[23]

In *One Native Life*, Richard Wagamese goes on a walk and sees a doe. In the meditation that follows, he realizes that "we don't become more by living with the land. Instead, we become our proper size. It takes unity to do that. It takes the recognition of the community we live in. This world. This earth. This planet."[24] These inquiries might not instantly move the thinker from ignorance and avoidance to purpose, but each person does move to a broader conception of themselves as part of a physical world of interconnections. Therefore, teachers looking to add depth to students' understanding of their identity within the physical world should build thematic inquiry units organized around appropriate questions that lead to credible conclusions.

An inquiry approach offers several advantages. First, it offers a necessary interdisciplinary flexibility that ecocriticism often demands. As Garrard explains,

> Environmental problems require analysis in cultural as well as scientific terms, because they are the outcome of an interaction between ecological knowledge of nature and its cultural inflection. This will involve interdisciplinary scholarship

> that draws on literary and cultural theory, philosophy, sociology, psychology and environmental history, as well as ecology.[25]

To this list, I would add economics, political science, agriculture, and urban design. The necessity of interdisciplinary flexibility is also reflected in the way that writers attempt to explain the climate crisis to a popular audience. In *The Discovery of Global Warming*, Weart offers a history of science. He chronicles the work of climate scientists as they reached a consensus around anthropogenic global warming. In *Don't Even Think About It*, George Marshall explores rhetoric and psychology as he explains why it is so difficult for people to change their lives in response to climate science. In *Dirt to Soil*, Gabe Brown explains how he changed his farm to restore nutrients to his soil without relying on synthetic inputs. Any interrogation of planetary systems and our connection to them will be interdisciplinary. As Bill McKibben points out in *Falter: Has the Human Game Begun to Play Itself Out?*, "*Everything* comes with strings attached, and you can follow those strings into every corner of our past and present." Emphasis in original.[26]

Second, inquiry is grounded in student curiosity, which can do more to change our thinking than directly didactic approaches. There are many understandable reasons why people, including students, can find a study of the physical world disturbing or uncomfortable, and these feelings can incentivize motivated reasoning. In *Everybody Lies*, Seth Stephens-Davidowitz examines online searches in response to political speeches, and the results are often disturbing. When President Obama delivered a didactic speech about inclusion of Muslims, Google recorded more racist searches in the United States. The most effective way to cut through this bigotry, Stephens-Davidowitz found, was inspiring curiosity. He concludes that "subtly provoking people's curiosity, giving new information, and offering new images of the group that is stoking their rage may turn their thoughts in different, more positive directions."[27] Ultimately, my goal is to help students add depth and complexity to their understanding of the relationship between them and the physical environment, something that will require some students to overcome defensive thinking and motivated reasoning. Therefore, starting with questions makes sense if the goal is to interrogate our mental model for engaging with the physical world.

Students in my class therefore pursue their own lines of inquiry, ultimately resulting in a research essay. The topics are wide-ranging and interesting. Students will explore to what extent divestment, geoengineering, or climate policies can help to mitigate global warming. Once introduced to mind/body dualism, they often want to explore ways the physical world affects mental health (To what extent is the mind independent from the body?), and therefore they often want to study forest bathing or even how antidepressants affect mood. These projects always start with questions on a topic the student is curious about. Paul Hawken's *Drawdown*, which attempts to rank climate policies and technologies, can be a wonderful generator of inquiry. The range of options surprises (and often inspires) students, as does Hawken's sometimes counterintuitive ranking. While students engage in inquiry, I intermittently offer several lessons in which I strive to model inquiry for the class. These lessons seek to reveal, explore, and evaluate our anthropocentric ethos.

My "person versus nature" unit starts by asking what is "natural." Almost all my students carry an intuitive definition of "natural," so complicating it at the outset can help to establish the potential value of the unit. Their initial definitions and examples of "nature" typically include descriptions of temperate rainforests. When they think of things that are "unnatural," they think of things created or destroyed by human activity—a forest is natural, but cities are not. In "Artificial Wilderness," a chapter in *1491*, Charles C. Mann explains how indigenous communities planted orchards within forests. Today, they look like "natural" places, but they were somewhat designed by people. Conservation parks that are often viewed as "natural" tend to also have policies to stop "naturally" occurring forest fires, so the forests in those parks age unnaturally. Planted forests often seem "natural" from a distance while gardens and monoculture crops seem the height of artifice. Today the notion of "natural" is especially fraught. In *The End of Nature*, Bill McKibben argues that there is no untouched "natural world" remaining because "we have changed the atmosphere, and that will change the weather. The temperature and rainfall are no longer to be entirely the work of some separate, uncivilizable force, but instead in part a product of our habits, our economies, our ways of life."[28] Students often respond enthusiastically to this lesson because they like thinking "philosophically." For some students, it is very difficult to revise their initial assumptions about what is "natural," and it is worth engaging them in discussion about the assumptions people make, including how those assumptions lead to real-world ecological consequences. This discussion returns us once again to Crutzen's "Anthropocene," a concept that carries with it an awareness of how deeply and profoundly human activity has changed the planet and its systems.

It seems odd to me that we managed to bring about so much prosperity with our boundless cornucopian model while also bringing about so much ecological damage that we discuss the "end of nature." Why is it so easy to think that people and their choices are walled off from the physical world? One explanation for this cognitive dissonance can be found in Northrop Frye's theory of "garrison mentality." Briefly, Northrop Frye suggested in a 1965 essay that European settlers built garrisons to protect themselves and their communities against the terrifying wilderness and that a "garrison mentality" remains an enduring foundation of how many Canadians view the environment. The garrisons, literal and metaphorical, Frye speaks of were built alongside colonization and the frontier myth of endless resources waiting for settlers[29] to exploit. Is the frontier so threatening and terrifying that it must be walled off, or is it a terrific and vast land waiting to be fenced so that it can be exploited and developed? (Maybe neither.) These ideas are tied up in complex ways, so it is useful to explore them in more accessible texts. I typically juxtapose "It's Mine! Canada—The Right Land for the Right Man. Canadian National Railways—The Right Way!" against Shaun Tan's fantastic children's story, *The Rabbits*, which also allegorically explores colonialism, the frontier myth, and anthropocentric cornucopianism. Like the colonial settlers, the rabbits build walled cities (Frye might call them metaphorical garrisons) to separate them from the environment and to cut into the frontier's landscapes as they build their rabbit-centered civilization. They assign some of the land for waste, which is walled off from the city. Just as the rabbits don't see the "strings" that attach them to the physical world, my students typically do not conceive of the walls between them and the environment. These connections can be made visible by explaining to students how air pollution lodges itself in our lungs or by showing pictures of the Pacific Garbage Patch. Rather than imagining tiny, forever decomposing plastics, my students mostly imagine whole soda bottles in the ocean. They rarely realize that microplastics and air pollution end up in human bodies because they assume people are "walled off" from this undesirable matter. Finally, like many adults, they cannot explain in much detail where human waste goes (though in this last case I don't show pictures).

This content is challenging for students, and it leads to depressing thoughts about humanity. Are we, like Agent Smith suggests in *The Matrix*, a virus on the planet? This interpretation of humanity is perhaps as far as we can come from Michelangelo's depiction of humanity in *The Creation of Adam*. If we are as gods, as Stewart Brand has suggested, then we seem to be callous ones. If we are the sole holders of consciousness, and therefore uniquely capable of assigning value to things, why do we so often act only in our interests? To grapple with this question, students are asked to examine charts, graphs, and infographics from NASA, the NOAA, and Our World in Data that draw attention to some of humanity's less admirable environmental impacts. For some students, this will be the first time they've confronted and considered atmospheric concentrations of carbon dioxide, ocean acidification measurements, or melting ice. Aside from Agent Smith (and perhaps Charles C. Mann in "The State of the Species"), people rarely discuss humanity's domination of the planet as the act of a virus. But scientists do discuss invasive species. These ideas are tied up in complex ways, so it is useful to explore them in more accessible texts. I typically show Population Education's YouTube video, "World Population," alongside Jennifer Klos's TED Ed video on invasive species, "The Threat of Invasive Species—Jennifer Klos." After introducing a variety of ecological concepts, Klos's video finally asks, "Are we [humans] an invasive species?" This is the lesson most likely to give students a sense of doom, though that does not mean it will inspire change or action. In some cases, it can be worth underlining the trade-off we've made in dominating the planet to grow our population to nearly eight billion people, but students will tend to begin applying existential questions to this broad look at history. What's it all been for?

It is easy to think that these are new thoughts, but poets and philosophers have worried over the impact of modernity for a long time. To what extent is industrial modernity compromising our ability to lead fulfilled lives? The Romantics, for example, looked at industrialization and worried that these changes were diminishing their capacity to live fully engaged lives. Caspar David Friedrich's *Wanderer Above the Sea of Fog* depicts a man atop a mountain, strong and grand. By the end of the century, Claude Monet's smoggy *Houses of Parliament* paintings would emphasize pollution in industrial London. Which landscape would you rather live in? For William Wordsworth, most likely the former. The octet of Wordsworth's "The World Is Too Much With Us" emphasizes that people have become disconnected from nature—"little we see in nature that is ours" and "we lay waste our powers"—while the sestet expresses a yearning for a more natural world: "Great god / I'd rather be a pagan in a creed outworn; / so might I, standing on this pleasant lea, / have glimpses that would make me less forlorn." The final images of ancient gods, Proteus and Triton, rising from the sea are grand, much like the view shown in *Wanderer Above the Sea of Fog*. For the Romantics, people experience awesome feelings in the wild, or serenity in pastoral landscapes, but these places are increasingly either polluted, ruined, or discarded by the reckless advance of the modern world. I can inspire Romantic nostalgia in my students by showing them headlines from *The New York Times'* annual "The Year in Climate" series, but they otherwise tend to be mostly or else exuberantly modern. They love cell phones and international travel and indoor climate control. They often love urban architecture, seeing something as wondrous in the skylines of Manhattan and Hong Kong as they do in the mountains depicted in *Wanderer Above a Sea of Fog*.

The Romantics were skeptical of cities, but people are nevertheless migrating to them. The final lesson, therefore, interrogates cities. In what ways can cities help us to live sustainable and fulfilled lives? In *The Story of More*, Hope Jahren explains the global migration of three billion people to the city within the context of her lifetime. Jahren emphasizes that

> cities are the very definition of *more*—hot spots of humanity visible from space. At night, they pulse with artificial light, and from above they resemble nothing

> so much as the network of nerves in the brain. Gleaming dendrites of suburbs splay from the blazing nuclei of city centers, each one connected to its neighbor by an axon of glittering highway.[30] Emphasis in original.

Environmental discourse has often criticized cities. Joni Mitchell, for example, reminds us that they "paved paradise to put up a parking lot" in "Big Yellow Taxi." However, the trade-off is not so black and white. Jahren points out that

> the very poorest of the poor, the one billion people who struggle to survive on less than one dollar of spending power per day, are the people who were left behind in rural settings. [. . .] Every year, several million people move into urban slums because access to essential medical services and improved wages is still, as a rule, higher than the rural alternative.[31]

This point is echoed in other texts, including Stewart Brand's *Whole Earth Discipline* and Doug Saunders's *Arrival City*. However, the quickest way to demonstrate the appeal of cities, I find, is to simply ask students whether they plan to live in a city or in a rural area. They overwhelmingly are planning to live in cities and will cite social, cultural, and economic reasons to justify those plans.

Cities are dynamic and designed, though my students rarely realize either of those characteristics. In *Happy City*, Charles Montgomery distinguishes the land use seen in a photograph of Woodward Avenue, Detroit 1917 to the land use outlined in the 1937 World's Fair "Futurama" exhibit (designed by General Motors). The former image shows public transit, pedestrians, and some cars, whereas the second shows cars and roadways. For most students, what they understand of cities is represented in the Futurama exhibit. It is worth underlining that much of what we default to when imagining cities and progress is a near century-old model. Cities change, and cities as we imagine them did not just happen but were to some extent designed and then made a reality. Second, there are trade-offs within this design. From urban decay to sprawling suburbs, it is not hard to find complaints that much of the built environment, with its urban sprawl and its cookie cutter suburbs, is ugly. And unhappy. It is difficult to look at Edward Hopper's *Nighthawks* and not sense that cities offer a sort of isolation, even when the individual is not actually alone. Even if they want to live in cities, many students will say they associate them with poverty, income inequality, and pollution.

And yet, students also find cities exciting and "futuristic." They already carry in their minds two futuristic models: a sort of retro-futurism of cars, skyscrapers, and trains and a bleak dystopian cityscape, like the Los Angeles shown in *Blade Runner 2049*. Some students, especially if they want to go into architecture programs, imagine biophilic design, even if they have little understanding about gardening. Students who researched cities in their essays will mention building up rather than out, density, and the YIMBY (yes in my backyard) movement. Others will mention that they'd like to see more public transportation. It can be fun to explore ideas like walkability or biophilic design—or to read David Roberts's articles on Barcelona superblocks. To put it all together, we watch some of Gabe Klein's TEDx talk "Cars Almost Killed Our Cities, But Here's How We Can Bring Them Back." Students journal in response to prompts that invite them to consider the extent to which we today live in the dreams of people who lived nearly a century ago. What effects might our dreams today create in the future?

At this point in the unit, students have nearly finished their research essays, so I invite them to begin sharing the conclusions of their inquiry with each other as they advance through the revision process. It is always interesting to see the wide range of topics students choose to explore in their inquiries. Whether they explore the terraforming of extraterrestrial bodies or environmental policy or rewilding, the students are invariably exploring the relationship between people and the physical world. At the start of the unit, most cannot explain in much detail how the greenhouse effect works, to give one example, but this assignment requires that they dig deeper than statements like "climate change is an existential crisis." If they want to write about extinction and de-extinction, they will have to study biodiversity and track down Kolbert's *The Sixth Extinction*. If they want to write an essay entitled "Why I Hate Suburbs," they have to be able to explain the social and individual costs of that land use policy and will often track down even relatively obscure texts like Hurley's *Radical Suburbs*. If they are skeptical that reforms to urban land use can be brought about quickly enough to mitigate or adapt to climate change, they must cite evidence. More broadly, by sharing the wide range of topics they've studied, they help to reinforce that there are "strings attached to everything." This project also moves them away from hoping for a "silver bullet" solution to the climate crisis. I hope that they feel a sense of optimism and opportunity in the fact that there are so many things that can be and are being done, even if the larger project of reforming our environmental ethos is daunting.

In reviewing our learning at the end of the unit, we again return to our initial questions. Are people a part of the world or in some way separate from it? My students' reflections and conclusions are neither radical nor world breaking, but they are interesting. Many respond first to their experience within the unit as a learner. They're often proud of what they wrote, especially if they had to really "dig" in their research or "buckle down" in their writing. Often, they are simply intrigued by the subjects they studied, and I will never forget one student who praised the "beauty" of regenerative agriculture before his peers. Others are happy to have completed what they see as a "philosophical unit" and still others say uncomfortable things like "I'd rather do this unit 10 times than another Shakespeare play." I am pleased when my students move from climate change ignorance or avoidance to a sense of purpose, but some students double down on their priors, ranging from technological optimism to an anxious defense that they're "tired of this stuff." Some students will note that climate change remains a mostly depressing subject for them. There must be room in the class for all these statements, and I remind myself that I too sometimes feel despair about humanity's impact on the physical world. It can be useful to reassure students that their inquiry can still go deeper and that it's normal, and in fact admirable, if they are left with further questions to consider. Students are often producing their first reflections about their relationship with the physical world that are not founded on a "person versus nature" survival stories unit. And they will continue down different paths as they get on with the business of living in the developed world.

A thematic inquiry approach offers several advantages. First, English Language Arts has often been about identity and relationships, so an ecocritical exploration of what our identity means within the context of our relationship with the physical world makes sense within the discipline's traditions and is likely to feel less daunting to teachers of English. Second, "person versus nature" units have often implicitly or explicitly communicated that the relationship between people and the wilderness is combative, and that harm can only go in one direction (i.e., that the world imposes hardship upon the person) when it is hard to deny that people individually and collectively also pose problems for the planet and its systems. This unit allows for a more nuanced, complex, and realistic understanding of that relationship. Third, McKibben is correct that there are "strings attached to everything," so an inquiry model that allows students to follow their own inquiry into their relationship with the physical world broadens the study to include interdisciplinary sources. Finally, inquiries into this relationship almost always lead to considerations about what should be done. It's not enough to welcome students to "Club Climate Alarmed"[32] or to raise students' "Climate Freakout Level,"[33] to use Kimberly Nicholas's phrasing. When students say things like "I'm tired of talking about this," I interpret them as communicating that they don't want to dwell on no-win scenarios, and so we add a lot of value in helping them to dig into the details, devise solutions, and find opportunities to make a difference.

Most importantly, reading and writing are one way that we interrogate relationships, and our exploitation of the physical world's resources is too often taken for granted. Our anthropocentric ethos has allowed us to achieve many admirable goals, but we are now grappling with the limits and drawbacks of that framework. The English classroom can help to explore those limitations and to develop new ideas, especially if teachers are willing to rely on ecocritical ideas to interrogate our relationship with the physical world.

NOTES

1. "English Studies 12." Building Student Success—B.C. Curriculum. Province of British Columbia. Accessed 19 July 2021. https://curriculum.gov.bc.ca/curriculum/english-language-arts/12/english-studies.
2. Greg Garrard, *Ecocriticism* (New York: Routledge, 2012), 5.
3. Timothy Morton, *Being Ecological* (Cambridge: The MIT Press, 2018), XXX.
4. Garrard, *Ecocriticism*, 18.
5. Stewart Brand, *Whole Earth Catalogue, Whole Earth Catalogue: Access to Tools* (Fall, 1968), 2.
6. Martha C. Nussbaum, *The Cosmopolitan Tradition: A Noble But Flawed Ideal* (London: The Belknap Press of Harvard University Press, 2019), 2.
7. Nussbaum, *The Cosmopolitan Tradition: A Noble But Flawed Ideal,* 2.
8. Rutger Bregman, *Utopia for Realists: And How We Can Get There*, trans. Elizabeth Manton (London: Bloomsbury, 2017), 1.
9. Thomas Schelling, *Strategies of Commitment: And Other Essays* (Cambridge: Harvard University Press, 2006), 36.
10. Schelling, 54.
11. William D. Nordhaus, *Climate Casino: Risk, Uncertainty, and Economics for a Warming World* (New Haven: Yale University Press, 2013), 146.
12. Nussbaum, *The Cosmopolitan Tradition*, 16.
13. Nicholas, *Under the Sky We Make*, 8.
14. Garrard, *Ecocriticism*, 6.
15. Garrard, *Ecocriticism*, 21.
16. Garrard, *Ecocriticism*, 24.
17. Greta Thunberg, *No One Is Too Small to Make A Difference* (UK: Penguin Books, 2019), 7–8.
18. Kimberly Nicholas, *Under the Sky We Make* (New York: G.P. Putnam's Sons, 2021), 102.
19. Klein, *This Changes Everything* (Toronto: Knopf Canada, 2014), 3.
20. Klein, *This Changes Everything*, 3.
21. Klein, *This Changes Everything*, 7.
22. Michael Pollan, *The Omnivore's Dilemma: A Natural History of Four Meals* (New York. 2006), 6.
23. Mark Bittman, *Animal, Vegetable, Junk: A History of Food from Sustainable to Suicidal* (New York: Houghton Mifflin Harcourt, 2021), 50.
24. Wagamese, *One Native Life* (Vancouver: Douglas & McIntyre, 2008), 217.
25. Garrard, *Ecocriticism*, 16.
26. Bill McKibben, *Falter: Has the Human Game Begun to Play Itself Out?* (New York: Henry Holt and Company, 2019), 10.

27. Seth Stephens-Davidowitz, *Everybody Lies* (New York: Del Street Books, 2017), 162.

28. Bill McKibben, "The End of Nature," in *American Earth: Environmental Writing Since Thoreau*, Ed. Bill McKibben (The Library of America, 2008), 718–719.

29. Adam J. Barker and Emma Battell Lowmann. *Settler: Identity and Colonialism in 21st Century Canada* (Halifax: Fernwood Publishing, 2015).

30. Hope Jahren, *The Story of More: How We Got to Climate Change and Where We Go From Here* (New York: Vintage, 2020), 22.

31. Jahren, *The Story of More: How We Got to Climate Change and Where We Go From Here*, 23.

32. Nicholas, *Under the Sky We Make*, 34.

33. Nicholas, *Under the Sky We Make*, 35.

REFERENCES

Asafu-Adjaye, John, et al. "An Ecomodernist Manifesto." April, 2015. http://www.ecomodernism.org/ (Accessed February 21, 2021).

Barker, Adam J., and Emma Battell Lowmann. *Settler: Identity and Colonialism in 21st Century Canada*. Halifax: Fernwood Publishing, 2015.

Beach, Richard, et al. *Teaching Climate Change to Adolescents: Reading, Writing, and Making a Difference*. London: Routledge, 2017.

Bel Geddis, Norman. *Magic Motorways*. New York: Random House, 1940.

Bittman, Mark. *Animal, Vegetable, Junk: A History of Food from Sustainable to Suicidal*. New York: Houghton Mifflin Harcourt, 2021.

Brand, Stuart. *Whole Earth Catalogue: Access to Tools*. Fall, 1968.

Brand, Stuart. *Whole Earth Discipline: An Ecopragmatist Manifesto*. New York: Viking Adult, 2009.

Bregman, Rutger. *Utopia for Realists: And How We Can Get There*. Translated by Elizabeth Manton. London: Bloomsbury, 2017.

Brown, Gabe. *Dirt to Soil: One Family's Journey Into Regenerative Agriculture*. White River Junction: Chelsea Green Publishing, 2018.

Canadian National Railways. "It's Mine! Canada—The Right Land for the Right Man. Canadian National Railways—The Right Way!" Ca. 1920–1925, poster. Alberta on Record. https://albertaonrecord.ca/is-glen-1338

"English Studies 12." Building Student Success–B.C. Curriculum. Province of British Columbia. Accessed 19 July 2021. https://curriculum.gov.bc.ca/curriculum/english-language-arts/12/english-studies.

Frye, Northrop. *The Bush Garden: Essays on the Canadian Imagination*. Toronto: A List, 1971.

Garrard, Greg. *Ecocriticism*. New York: Routledge, 2012.

Harvey, Hal. *Designing Climate Policy Solutions: A Policy Guide for Low-Carbon Energy*. London: Island Press, 2018.

Hawken, Paul. *Drawdown: The Most Comprehensive Plan Ever Proposed to Reverse Global Warming*. New York: Penguin Books, 2017.

Hurley, Amanda Kolson. *Radical Suburbs: Experimental Living on the Fringes of the American City*. Cleveland: Belt Publishing, 2019.

Jahren, Hope. *The Story of More*: *How We Got to Climate Change and Where We Go From Here*. New York: Vintage, 2020.

Klein, Gabe. 2017. "Cars Almost Killed Our Cities, But Here's How We Can Bring Them Back." Filmed October 2016 in Washington, DC. TED video, 19:31, https://www.youtube.com/watch?v=GXn2Iu8mSWY.

Klein, Naomi. *This Changes Everything: Capitalism vs. the Climate*. Toronto: Knopf Canada, 2014.

Klos, Jennifer. "The Threat of Invasive Species—Jennifer Klos." TED-Ed. YouTube video, 4:45, May 3, 2016, https://www.youtube.com/watch?v=spTWwqVP_2s.

Kolbert, Elizabeth. *The Sixth Extinction: An Unnatural History*. New York: Henry Holt and Company, 2014.

MacKenzie, Trevor. *Dive Into Inquiry: Amplify Learning and Empower Student Voice*. Irvine: Edtechteam Press, 2016.

Mann, Charles C. *1491: New Revelations of the Americas Before Columbus*. New York: Vintage, 2005.

Mann, Charles C. "State of the Species: Does Success Spell Doom for Homo Sapiens?" *Orion Magazine*. https://orionmagazine.org/article/state-of-the-species/ (Accessed January 26, 2021).

Marshall, George. *Don't Even Think About It: Why Our Brains Are Wired to Ignore Climate Change*. New York: Bloomsbury. Publishing, 2014.

McKibben, Bill. "The End of Nature." *American Earth: Environmental Writing Since Thoreau*, edited by Bill McKibben, The Library of America, 2008, pp. 718–24.

McKibben, Bill. *Falter: Has the Human Game Begun to Play Itself Out?* New York: Henry Holt and Company, 2019.

Montgomery, Charles. *Happy City: Transforming Our Lives Through Urban Design*. Toronto Doubleday Canada, 2013.

Morton, Timothy. *Being Ecological*. Cambridge: The MIT Press, 2018.

Nicholas, Kimberly. *Under the Sky We Make: How to Be Human in a Warming World*. New York: G.P. Putnam's Sons, 2021.

Nordhaus, William D. *Climate Casino: Risk, Uncertainty, and Economics for a Warming World*. New Haven: Yale University Press, 2013.

Nussbaum, Martha C. *The Cosmopolitan Tradition: A Noble But Flawed Ideal*. London: The Belknap Press of Harvard University Press, 2019.

Pollan, Michael. *The Omnivore's Dilemma: A Natural History of Four Meals*. New York: 2006.

Roberts, David. "Barcelona's Radical Plan to Take Back Streets From Cars." Vox.com. https://www.vox.com/energy-and-environment/2019/4/9/18300797/barcelona-spain-superblocks-urban-plan (Accessed January 26, 2021).

Rosling, Hans. *Factfulness*: *Ten Reasons We're Wrong About the World—and Why Things Are Better Than You Think*. New York: Flatiron Books, 2018.

Saunders, Doug. *Arrival City: The Final Migration and Our Next World.* Toronto: Knopf Canada, 2010.
Schelling, Thomas. *Strategies of Commitment: and Other Essays.* Cambridge: Harvard University Press, 2006.
Stephens-Davidowitz, Seth. *Everybody Lies: Big Data, New Data, and What the Internet Can Tell Us About Who We Really Are.* New York: Del Street Books, 2017.
Thunberg, Greta. *No One Is Too Small To Make A Difference.* UK: Penguin Books, 2019.
Wagamese, Richard. *One Native Life.* Vancouver: Douglas & McIntyre, 2008.
Williams, Florence. *The Nature Fix: Why Nature Makes Us Happier, Healthier, and More Creative.* New York: W.W. Norton Company, 2017.

Chapter 5

Foregrounding the Value of Ecocriticism in a South African University Context

David Robinson

In September 2017 I flew from Johannesburg to Cape Town in order to attend an academic conference. Prior to departure I had read news reports of the water crisis in Cape Town. After I collected my luggage and walked through the airport to the taxi rank, I became aware of the scale of the problem. Written on the airport walls, in letters that were six-foot high, were public service messages like "Cape Town is experiencing a water crisis. Use water sparingly." On arriving at my hotel, the scale of the problem was emphasized—the hotel staff requested that guests should not bathe, but rather shower, because showering uses less water. The request was that showering should be limited to two minutes per person per day. The city was dealing with a drought, brought about by a lack of rain. This lived experience, of dealing with a water supply problem, made me recognize that the planet was undergoing changes in climate and that humans were becoming more aware of a need to think more effectively about environmental concerns and to modify our behavior.

A few years prior to the conference in Cape Town I had made the decision to introduce ecocriticism as a topic of study in the Third Year Bachelor of Education module on teaching methodology in English studies at the University of Johannesburg (UJ), in South Africa. I was the coordinator of the module and together with another colleague presented the classes to the students. My focus was the study and teaching of English literature, and I became aware of the discipline of ecocriticism through my own reading and research. The decision to include the topic was made on the basis that within the concept of ecocriticism is a significant and valid set of theories to study at university at this point in human history. The methodology prepares English teachers for their role in South African secondary/high schools.

EDUCATION POLICY AND THE NATIONAL CURRICULUM DOCUMENTS

The choice to include ecocriticism was based partly on the policy documents for Third Year Methodologies in the Faculty of Education at the UJ. One of the topics informing this set of documents was the concept of Big Ideas. In faculty discussion it was agreed that the concept of Big Ideas could be interpreted as core concepts in a discipline. For the English Methodology module, the core concepts were drawn from the existing national Curriculum and Assessment Policy Statement, referred to as CAPS.[1] These core concepts included the traditional elements—writing, grammar, listening and speaking, and literature. The CAPS document, published in 2012, was a development of the previous national curriculum document, the National Curriculum Statement of 2008.[2] Both documents foregrounded the need to address matters of social and environmental justice. Using these broad concepts as a basis for teaching literature I chose to include ecocriticism as a topic.

The two curriculum documents make statements about human rights and social and environmental justice; they also provide a commentary on why these are valuable aspects that underpin the school curriculum. The 2008 NCS states the following:

> Literacy (reading, writing and viewing) provides access to information, lifelong learning and work opportunities. It is also an important tool to understand and assert one's human rights. Therefore, it is a key aspect of social justice. Schools must enable learners to achieve high levels of literacy in order to develop an understanding of social and environmental justice.[3]

The issues of human rights and social justice are matters that are foregrounded in the South African curriculum. Bearing in mind the country's history, in which a previous oppressive regime practiced discrimination on the basis of race (and also on the basis of gender), these matters are significant and important to acknowledge and address. Since the advent of the democratic government in 1994 various social institutions, including education, have made an attempt to address the social inequalities that still exist.

From the same curriculum document these issues are commented on as follows:

> By enabling learners to acquire critical literacy, teachers make an important contribution to social and environmental justice. Therefore, they should choose texts that enable learners to engage critically with their world, for example texts about access to the resources in our society (health services, water, wealth, etc.)

> and the way in which these resources are used. These texts will support the themes used for teaching and learning.[4]

The idea of environmental justice includes access to resources as well as the matter of ensuring that the environment can be sustained. This aspect of the curriculum directly addresses the topic of climate change, as well as the need to provide knowledge and debate on the matter of ecological thinking.

Similar ideas are echoed in the 2012 document, the CAPS document, which is currently used in South African schools. It addresses the issues in this way:

> Human rights, inclusivity, environmental and social justice: infusing the principles and practices of social and environmental justice and human rights as defined in the Constitution of the Republic of South Africa. The National Curriculum Statement Grades R-12 is sensitive to issues of diversity such as poverty, inequality, race, gender, language, age, disability and other factors.

This statement is similar to that of the previous curriculum document; there is more detail provided, although the matter of environmental justice is referred to but not explored in detail.

The CAPS document makes several statements about the teaching and learning of literature, including this focus:

> The main reason for reading literature in the classroom is to develop in learners a sensitivity to a special use of language that is more refined, literary, figurative, symbolic, and deeply meaningful than much of what else they may read. While most literary texts are forms of entertainment, amusement, or revelation, serious writers create novels, plays, and poems because they have ideas, thoughts, and issues, and principles, ideologies, and beliefs that they most want to share with or reveal to their prospective readers. Their imaginative use of language is an added method of revealing, reinforcing, and highlighting their ideas.[5]

This is a broad statement that covers a range of matters. There is no specific link between this statement and the matters covered in the document with regard to social and environmental justice. However, the references to principles and ideologies are probably linked to these matters without being foregrounded. In choosing to prepare preservice teachers to study literature through an ecocritical lens, we connect the value placed on "serious writers" who bear witness to social and environmental issues to the objective of preparing "learners to engage critically with their world."

The CAPS document states the following about literary criticism:

> Literary interpretation is essentially a university-level activity, and learners in this phase do not have to learn this advanced level of interpretation. However, the whole purpose of teaching literary texts is to show learners how their Home Language can be used with subtlety, intelligence, imagination, and flair. This means taking a close look at how a text is being created, manipulated, and re-arranged to clarify and emphasise what is being expressed.[6]

The ecocritical perspective can be accommodated in the curriculum because of this potential for flexible teaching and learning approaches. Indeed, based on the underlying principles presented in the extracts above, ecocriticism is a concept that is an effective fit with the requirements of environmental justice.

THE LIVED EXPERIENCE OF SOUTH AFRICANS AND THE MATTER OF THE ENVIRONMENT

There are several challenges that exist in the South African environment. This is an important matter because the lived experiences of the students will inform, to some degree, how they will respond to the material and concepts presented to them. A range of ecological matters have a role in the lives of South African students and the society at large.

A key issue is the water crisis that is most well-documented in Cape Town but has also been noted in other parts of the country. In Johannesburg, the most populous city in the country, and its economic hub, there is an ongoing set of water restrictions in place. The rainy season for this city tends to be in the summer months (September to March, different from the Northern Hemisphere). During these months there is rain and the local dams tend to be filled; these then supply water to the city and the surrounding areas for the autumn and winter, during which little rain falls. The dam levels tend to drop during the winter months and, to sustain the supply, restrictions on water usage are in place. These restrictions include that gardens cannot be watered during the day, but only after 6:00 p.m. and before 6:00 a.m. Swimming pools may not be filled with municipal water supply. Cars may not be washed with hoses—buckets of water are regarded as fit for this purpose. These measures have reduced the usage of water in the city.

In 2017 and 2018 the city of Cape Town was required to put in place measures to ensure that the water supply to the city was not totally depleted. In this city, during the worst period of drought, there was a limit put on the number of gallons of water each citizen could/should use per day. The journalist Aryn Baker, writing in early 2018, provided several details of the dire circumstances in which the city was operating.

Before commenting on the city's actions, Baker makes this point about the reasons for the drought:

> As with the California dry spell [mentioned elsewhere in the Baker article], climatologists at the University of Cape Town say manmade global warming is a likely factor in the continued drought and that we, like many other cities around the globe, are facing a drier future with increasingly unpredictable rains.[7]

This point is a significant one because it states the possibility that citizens of other cities could find themselves in a similar situation in the future. The requirement for one city to address a reduced water supply is something from which other city managers can learn. In terms of the restrictions put in place, Baker states,

> We are now limited to using 13 gallons of water per person per day. That's enough for a 90-second shower, a half-gallon of drinking water, a sinkful to hand-wash dishes or laundry, one cooked meal, two hand washings, two teeth brushings and one toilet flush.[8]

Baker also comments that one of the five dams that supplies Cape Town with water was at 13% capacity on February 13, 2018.[9] The other dams were also in a state of reduced water levels.

The city was faced with the possibility of what was termed "Day Zero"—the point at which the water supply to the city would be totally depleted, and which would result in various other limited water supplies (the delivery of bottled water being one such possibility). As things turned out, the point of total depletion of water supply fortunately did not occur. The management of the available water through restricted usage provided the city with the time to get to the next rainy season and the replenishment of the dams. Disaster was averted through a change in human behavior.

Apart from the restrictions on water in Johannesburg and Cape Town, there are other places in the country where there is a water supply problem. The period of drought was broken by the return of seasonal rains (in Cape Town the rain normally takes place in winter months, unlike the majority of the country). However, it is expected that this challenge will be a repeated problem in the coming years.

Apart from the issue of interrupted rainfall, there is another matter related to climate change that has an effect on the coast. This factor is the effect of flooding on the value of private residential property on the South African coastline. Andre Kruger, in his doctoral thesis of 2019, addresses this matter using an established understanding of the term "climate change":

> The most commonly used definition is that of the UNFCCC (1992:7), which defines climate change as "a change of climate which is attributed directly or indirectly to human activity that alters the composition of the global atmosphere and which is in addition to natural climate variability observed over comparable time periods."[10]

Kruger's statement addresses global matters relating to climate change and notes that there are South African authors who address these issues at the local level. Several authors have commented on the potentially negative effects: "An increase in sea level, wind velocity, storms, altered freshwater flows and shoreline erosion were identified as some of the potential consequences of climate change."[11] Apart from the negative consequences for property values (Kruger's main research interest), there is a concomitant effect on the well-being of the people, wildlife, and flora in these coastal situations. Kruger points out that rising sea levels, resulting from global warming, will have a negative effect on estuaries, wetlands, inlets, and marinas on the South African coastline.[12]

Kruger states that the decision to buy residential property is a significant decision for most buyers, because the property becomes a home with emotional and personal attachments.[13] He adds that the decision is also informed by weather events "such as the storm swell on the South African KwaZulu-Natal coast in March 2007. This event caused physical damage and behaviour changes in the South African coastal residential real estate market."[14] Kruger makes the point that, as a result of the storm swell along the KwaZulu-Natal coastline, property prices in the area fell. This has a material consequence for people whose personal wealth is compromised because of damage to property. Kruger also addresses wider considerations regarding the consequences of climate change, including land use such as urbanization, which can have a negative impact on ecosystems.[15]

Apart from the issues of water supply and damage to coastal properties, there are other matters that are also part of the ecocritical debate, and these include matters relating to the use and abuse of animal resources. There are many examples of concern relating to wildlife in South Africa. These include the poaching of animals such as rhino and elephant. These matters have been part of the public discourse for some years. In the preface to her book on the threats facing the survival of elephants and rhino, Bridget Martin makes the statement that

> poaching and illegal trade in ivory and rhino horn have caused their [elephants and rhino] numbers to plummet, and it is debatable whether sustainable use in these commodities is now viable as some species and populations are teetering on the brink of extinction.[16]

A clear indication of the dire circumstances in which these animals exist, Martin adds that in 2011 the BBC reported that rhino horn was worth more than gold or cocaine in terms of the amount paid per kilogram. This amount was approximately $65,000 per kilogram.[17] This financial incentive drives the poaching industry and puts the lives of these animals in danger. Martin makes the point that while there are people willing to make a profit from the killing of these animals, there are also people who are willing to protect the threatened species.[18]

In addressing this matter it is important to recognize that there are cultural beliefs and norms involved in the trade of animal parts. Cultural beliefs frame the social existence of people, and they play an important role in the creation of values that govern behavior. They are also central to the functioning of a society and are deep-seated, regarded as inherently valuable. However, it must be acknowledged that cultural beliefs are not cast in stone—they are human creations and cultures change over time, sometimes because of internal forces, and sometimes because of external forces. Through educational engagement it is possible to suggest different values and actions.

Another element relating to the issue of animals is the specific and selective breeding so as to create rare or unusual variants. An article in *The Mail and Guardian* weekly newspaper commented that "from 2009 to 2014, prices for exotic game shot through the roof. The selling price of sable [antelope] rose by 479% and of disease-free buffalo by 540%, according to figures from Wildlife Ranching South Africa."[19] The market began to collapse in 2015, and now breeders of exotic animals, including color variants, are not able to recoup their initial payments.[20]

The breeding of exotic animals is an example of the bizarre choices made by some of the wealthy in South African society. Why anybody would desire to possess a golden wildebeest is not an easily answerable question. Possibly the desire derives from the need to own that which is rare, even if it is unnatural. There are elements to these breeding practices that have a hint of Dr. Moreau to them, as well as Dr. Frankenstein. These animals exist because somebody chose to manipulate a natural status for their personal benefit. This type of action raises questions about what the next step will be: Will there be a desire for polka-dotted animals?

Although the issue of the breeding of rare animals is not directly linked to climate change, there is an underlying attitude that is connected with both practices: the attitude that human beings have dominion over the earth and everything on it. In my classes I address this matter, and engage with the idea of the nature of human beings. It is noted that some people believe that human beings are more significant than animals because they are described as being created by God in his image. Other people believe that human beings are more significant because they have an intellectual ability that sets them apart. However, these arguments were used at various times to discriminate against other groups of human beings. In South Africa during the apartheid era the vast majority of the population (black Africans) were provided with a more limited education because they were regarded as less intellectually capable as a group.

Another matter to consider is the availability and use of mineral resources, in this case the use of coal for South Africa's power stations. The vast majority of the country's electricity is created and managed by the company named Eskom (which stands for Electricity Supply Commission). This company uses coal-fired power stations to generate the majority of the country's power supply.

Michael Bloomberg and Carl Pope make reference to the damage that coal causes to the environment and argue that there are more environmentally friendly options to generating power.[21] The renowned British naturalist Sir David Attenborough makes the point that there are alternatives to fossil fuels. He states that at least three countries are no longer reliant on fossil fuels—Iceland, Albania, and Paraguay.[22] These countries are not the main economic powerhouses on the planet, yet they have made a decision to move to electricity supply that is cleaner and better for all life. On this topic, Gaia Vince states, "Decarbonizing is hard. Access to energy is an essential part of development, and around the world 1.3 billion people lack electricity." Without electricity the standard of living of the people is reduced.

Fossil fuels are one way to provide this energy. Attenborough makes an additional point about how electricity supply has changed in some countries. He identifies Morocco, in North Africa, as a country which at one point generated almost all of its energy through imported fossil fuels. Now, however, Morocco generates 40% of its energy from renewable sources, including the world's biggest solar farm.[23] He adds that there is a possible future in which Morocco could become a net exporter of electricity, through undersea cables to the nearby European continent. This is an example of what can be achieved through innovation that does not rely on fossil fuels.

At this point there is little evidence that the South African electricity supply will move away from coal to a significant degree in the near future, although there is one nuclear power station at Koeberg. Because of the unstable power supply, many South African citizens are choosing to install their own solar power panels for domestic use. At UJ the solar power initiative has been recognized; one of the larger car parks, which was previously uncovered, has been turned into a covered car park with solar panels across the whole surface. Students and lecturers who attend classes are part of a physically changed environment in which clean power is being provided, although the solar power does not address all of the needs of the institution. For the people at UJ the idea of solar power is a lived reality, not just a concept in a textbook, or in images on a screen.

THE THEORETICAL PERSPECTIVES

The students in the class were undergraduate students in their third year of a four-year degree. They were preservice teachers and drew a lot of their knowledge from the English Department, which addressed the teaching of literature as a subject. The students in my class were studying to become English teachers, and they were not aware of any ecocritical theories at the start of the contact sessions. My intention was to provide the students with an extension of their existing knowledge, which included literary knowledge of postcolonial theory/ies, feminism, and close reading. I wanted them to realize that it is possible to approach literary texts in a way that foregrounds ecocritical concerns. In the contact sessions I provided students with definitions of the concept, as well as a brief historical statement about the concept, and some basic points of reference regarding the positions that are espoused by various ecocritical thinkers.

Following the definitions and the theoretical positions, examples of literature were presented so that students gained an understanding of the application of theory. Part of the process was to engage with students in an online discussion forum on the university platform. The final element of the teaching and learning process was that the students were provided with an assessment opportunity in which they could engage with the concept. They were each required to provide reasons regarding the value of ecocriticism in the English class, as well as an example of a text that they believed was a useful resource for addressing ecocritical thinking; further, they needed to indicate the approach/strategies they would use in the teaching process.

The definition that is used in this module is that provided by Cheryll Glotfelty: "Simply put, ecocriticism is the study of the relationship between literature and the physical environment. . . . Ecocriticism takes an earth-centred approach to literary studies."[24] This definition is of value because it provides a succinct statement about this literary concept. This position is different from that which is covered in the English major, in which theories foregrounding postcolonial thinking and an exploration of gender-related theory tend to be the basis for engagement. Because ecocriticism is not foregrounded in the English major modules there is a need to address the theories of ecocriticism in the Methodology module.

In addressing the theoretical aspects of ecocriticism we include reference to the history of the concept; Timothy Clark makes the point that the discipline is a relatively recent phenomenon, and he states that as a "defined intellectual movement it is largely dateable to the founding of the Association for the Study of Literature and the Environment in 1992."[25]

Clark's statement identifies ecocriticism as a fairly recent academic discipline, but Clark and other critics such as Greg Garrard make the point that some ideas in the discipline were posited by authors such as Rachel Carson in the early 1960s. The formal development of the theories informing the discipline, however, took place more recently. The more recent commentators have critiqued the discipline itself as well as refining the concepts presented by earlier thinkers in the field.

To introduce the discipline of ecocriticism to the class, the students were provided with questions to consider as a way to approach it. The questions were intended to focus the students' attention on matters such as the representation of nature in literature, how different genders write about nature or the physical environment, and how the physical setting supports the reader's understanding of a text. In addition, the questions directed the students' attention to the way in which figurative language was used in literary texts in relation to the environment.

The theoretical work of Greg Garrard is a central feature of the Methodology module. Some elements of Garrard's 2012 text[26] are used as organizing elements that provide a set of theoretical positions in the discipline. From Garrard's commentary several theoretical positions are foregrounded in the module: cornucopia, environmentalism, ecofeminism, and deep ecology. Students are provided with information on other positions, such as Marxist ecocriticism, but these are not foregrounded in the same way.

One position is that presented by the cornucopians. A significant criticism of this group is that it takes little or no account of the nonhuman environment except insofar as it affects human wealth or welfare: "Nature is only valued in terms of its usefulness to us."[27] In his commentary Garrard makes the point that this is not really a theory addressing the environment in terms of concern for the environment, but rather in terms of its use for the human population. In this sense, the position is not really an environmental position but rather one about resources.

The second position is that of environmentalism. Garrard states that this applies to a broad range of people who are concerned about environmental matters such as global warming, but who would wish to maintain their standard of living. They are concerned about pollution and scarcity of natural resources, but they look to their governments or nongovernmental agencies to provide solutions.[28]

The third position the module addresses is deep ecology. This position states that the well-being and flourishing of human and nonhuman life on Earth have value in themselves (in other words, all life has intrinsic value and inherent worth). These values are independent of the usefulness of the nonhuman world for human purposes. This position was proposed and developed by the Norwegian philosopher Arne Naess.[29] There are aspects to the position that reconfigure the relationship between humans and the environment, compared with other common positions in which the desires of human beings are regarded as most valid.

Finally, the module addresses the theory of ecofeminism. The point is made that deep ecology identifies the anthropocentric dualism humanity/nature as the ultimate source of anti-ecological beliefs and practices, but ecofeminism also considers the androcentric dualism man/woman. Ecofeminism involves the recognition that these two arguments share a common "logic of domination," or underlying "master model," that "women have been associated with nature, the material, the emotional, and the particular, while men have been associated with culture, the nonmaterial, the rational, and the abstract."[30]

The four positions were chosen because they tend to fit on a continuum in which the degree of concern for the environment is ever more foregrounded. Cornucopians address the environment as a resource, whereas deep ecologists argue that the environment is far more than that, and not only human purposes have value. In addition to the four theoretical positions mentioned, the module uses Garrard's concepts of pollution, wilderness, apocalypse, and animals as points of interest. Based on the lived experiences of South Africans, the possibility of exhausting a water supply (leading to possible apocalypse), or hunting animals to extinction (animals are central to this matter), these four concepts have great significance. Garrard explores each in terms of the written/literary texts that inform these ideas. It is notable that in many cases the texts to which Garrard refers are nonfiction, addressing topics such as the consequences of overpopulation. However, in cases addressing topics such as apocalypse, there is also a reference to speculative fiction. Further, Garrard comments on Shakespeare's references to or use of nature in his plays. Examples include the influence on nature reflected in *A Midsummer Night's Dream*, in which natural order is compromised because the fairies, who care for nature, are in conflict. Another example of manipulation of nature occurs in *The Tempest*, in which Prospero uses his magical craft to influence the weather. A third example is linked to the disorder of nature that arises in *Macbeth* after the murder of Duncan: the king's horses eat each other, a clear example of disorder in nature.

The literary element of the module addresses the idea that human cultural expressions provide a different perspective from that of science; ideas are explored in a literary manner, or through specific genres, and these expressions address fears and concerns, but also a sense of enjoyment or celebration of human lives within an environmental space. A poem, or play, or novel, presents a different medium of expression from a scientific text. However, there is a literariness in some scientific writing—for example, Darwin's text on the voyage of the Beagle is regarded as a scientific account but also as literature.

One text for class discussion is J. M. Coetzee's *The Life and Times of Michael K*,[31] which we analyze certain aspects of based on identified ecocritical features. For example, Michael K, the main character, is a gardener.[32] A gardener is connected to natural processes because gardening work requires engagement with plants. The work includes providing plantlife with irrigation, as well as a location with appropriate sunlight, and the removal of parasites. These actions benefit the plants. However, gardeners act in other ways too. They decide on the composition of plants in the garden, selecting and rejecting particular types. In this way gardens are shaped to suit human purposes. In a well-mown lawn and a well-kept flower bed there is a sense of artificiality; if these plants were left to their own devices, the resulting garden would be far less orderly. Gardeners, therefore, are knowledgeable about plants, the seasons, the dangers of parasites, and the need to nurture plants. However, the purpose is a human one, in that the garden is designed and intended to provide a space in which human beings can feel fulfilled.

It is notable that Michael K is facially deformed: he has a harelip.[33] This marks him as the Other, in that he is different from other humans. He does not engage with people as others do. From birth to adulthood he is aware of the otherness of his appearance and how this locates him on the margins of human society. In a sense he becomes a gardener because other avenues are closed to him. Gardening is a valid occupation for somebody already removed from society because it can be done in relative solitude. When he changes work from gardening to working as a guard, he is the victim of violence (he is attacked at night) and returns to working for the city in the parks section, as a gardener.

In the central section of the novel Michael spends time on an abandoned farm which he believes is the place where his mother was born. His knowledge of gardening allows him to grow food for himself and to survive. (Coetzee is a vegetarian, and this aspect of the novel might be a personal statement of how human beings could live, as compared to the more extravagant usage of the land in the field of commercial agriculture.) The final sentence of the novel is a statement of living in a world without excess. The image is of a man using a spoon on a string to draw water from a well, and to then consider the spoonful of water, and to state that it is possible to live like this (i.e., on the water in the spoon).[34] The implication of a person being able to live on a spoonful of water is that of simplicity and minimalism. Coetzee suggests that we do not need all of the possessions that we desire in order to live in a way that is fulfilling. Symbolically, Coetzee suggests that we can live in a simpler and less complicated way. From an ecocritical perspective, this provides insights into acknowledging that we need to reduce waste, and also that our desires and our needs are not necessarily the same thing.

Another Coetzee text that is referenced, briefly, is *The Lives of Animals*.[35] The text addresses (among other things) the slaughter of animals in abattoirs and compares these deaths to the deaths of the Jewish people who were victims of the Holocaust in World War II. The text brings into question the morality of ordinary human actions, such as eating meat, and the value of human as opposed to nonhuman animal life. The text addresses highly complex ideas that have drawn comment from established academics in the fields of animal rights, human rights, and history. Students are asked to engage with these topics in online discussion boards.

A third novel that is considered is the Young Adult novel *Savannah 2116 AD* by Jenny Robson.[36] This is a South African novel that won the Percy FitzPatrick Award, awarded by The English Academy of Southern Africa. The novel is set in the future, in which select human beings are raised in institutions; their organs are donated to animals that are endangered or on the brink of extinction, to ensure the continued existence of the species. The novel raises questions about what humans should do to address the damage they have done to the environment. It also addresses the idea of a draconian government that uses military might to ensure that the population complies with the law.

The novels presented are all from South African authors (although Coetzee is now a resident in Adelaide, Australia). The texts were chosen as examples because they connect to some aspect of South African society. Coetzee's later novel, *Disgrace*, is more frequently taught in South Africa and includes some ecocritical aspects to it, such as who should have the ownership of farmland, and how animals should be treated.

Having explored the theoretical background to the module, I will now move on to explore the student responses to the assessment, which was a written assignment. Students were required to identify an ecocritical position and to choose a text that could be presented in school classes in such a way that the theory could be used as a lens to approach the text. They could choose any text that they felt was of value in teaching the topic and while several of them chose texts that had been discussed in class, many presented texts that had not been considered at all. The assignment provided direction about what was required but the key objective was to show how an aspect of theory connected to the text at hand and to share how their intended teaching strategies would help illustrate the connection.

STUDENT RESPONSES

In the class there were over 200 students, and they provided a wide range of responses.[37] A significant number of students chose to address the topic of ecofeminism and selected texts which they saw as valid to approach through this critical position. Notably, the majority of the class was female so it is perhaps not surprising that the majority chose a theory that articulates with their own circumstances.

The extracts below were taken from student responses to provide an indication of how the group approached the topic. Most provided a definition (taken from the notes) and then explored the ideas presented in the material. They then provided insights into why they believed a particular text was well-suited for instruction. One student observed that

> ecocriticism is important in teaching English in schools, especially for learners who reside in rural areas. This is because some rural areas do not have water taps. People rely on fetching water from the river. Therefore, ecocriticism in rural schools is important because learners would know the importance of keeping the water clean to prevent it from being polluted. They would also understand the importance of preserving water and use it responsibly.

This comment is of interest because it addresses the environmental differences between the urban and rural communities in the country. The assertion indicates that rural communities are more engaged with their water supply—they physically collect it from rivers, and they have a role in the collection and transportation of the water. Because of this they have a different relationship with this vital liquid than those of us who live in cities, suburbs, or towns. The experience of water supply in these larger communities is done through a third party, and there is less of a sense of personal responsibility for the provision for self, family, or others. In communities that are urbanized there is a sense that water, like many other elements, is commodified, and the liquid is managed and provided through an anonymous corporate entity. Water provided in city or suburban taps is put through a processing system to ensure that it is drinkable, but the process is unknown to the consumer. The effort of fetching water from a river is a more hands-on matter and the people engaged in this process have a greater awareness of water as a resource. They are likely to note when water levels drop, or when the river is likely to come down in flood. For these people the concept of a drought is not an abstraction. This experience has value in articulating the ideas contained within the ecocritical discipline. There are several literary texts that can be used to explore this matter of water as it connects to the relationship and difference between rural and urban communities.

A text that was not taught or mentioned in the module, but which has a link to the issue of water and human needs, is Walter Tevis's *The Man Who Fell to Earth*. The 1960s science fiction novel was made into a film directed by Nicolas Roeg and starring David Bowie. It tells the story of an alien who has come to Earth because his home planet has run out of water, and he seeks assistance in accessing water. However, it is an interesting footnote rather than a text that is suitable for the students of the class in which the ecocritical material is covered. A more accessible text is Alan Paton's *Cry, the Beloved Country*. This 1948 novel is set for Grade 12 students nationally, and it contains references to the difference between rural and urban settings, starting in the small village of Ixopo in Natal (now KwaZulu-Natal),[38] and including a memorable description of the train journey across the country to the city of Johannesburg. The restricted living conditions in the urban environment contrast significantly with the open savannah of the rural landscape. Notably, the first person that the protagonist meets in the city, after his journey, turns out to be a criminal who cheats him out of some money—setting the scene for the type of danger and risk associated with living in the city.

Another novel in which urban and rural communities are presented is Nadine Gordimer's *July's People*. In this text the setting is South Africa after a fictional revolution, and suburban dwellers flee to the rural community for safety. Again, there is a sense of the difference brought about by differences in environment.

Ecocriticism addresses issues about literature and the environment, and this student's comments do not reference literature—they are focused entirely on the lived experiences of rural communities. In a sense this response needs greater complexity—there is a need to address the literature connection. While students were required to address literary texts of various genres in their written statements, in some cases the literary commentary was fairly undeveloped.

Another student responded that

> in South Africa we face many environmental problems like air and water pollution. Many rural townships and communities do not have running water in 2020. They have to use water from the rivers and lakes to do basic everyday chores like cooking and cleaning.

This is another example of water supply to rural communities not being provided through pipes and taps, but rather from rivers and lakes. It is important to note that more than one student addressed this topic as it is clearly of significance. The water supply, or lack of it, can determine whether people live or die, and also have an effect on the fauna and flora. Notably, the matter is not simply an abstraction—it is part of the lived experiences of several of the students and their communities. The matter can be linked to the concept of environmental justice, in that one element of this topic would include access to resources in the environment in order for a community to subsist.

Another student noted that

> ecocriticism in the classroom is a topic that addresses how learners express, conceptualize and also how we deal with the environment. It is a consideration of how humans express themselves culturally specifically in literature, when dealing with issues relating to the environment.

This response is of value because it indicates that the student understands there is a relationship between the writer, the reader, and the environment, which is mediated through the written word. There is an implication that this type of engagement is not a mere record of events but holds within it the subjective experience and statements of the writer, as well as the exploration of these ideas by the reader. The location of the expressions about the environment within a cultural space is of significance, because there is an implication that values are expressed and explored, and that creative elements of expression are welcomed. As an extension of this there is the sense that ideas might be presented for debate or consideration, and that no single answer is necessarily sought.

Another student observed that

> as an English teacher, it is important to use literature in the classroom to explore the different issues of the environment. Learners need to be taught that they should not only look to the government to find solutions to environmental issues, but they should take responsibility too.

This statement supports the idea of critical thought as part of the literary engagements at school. In addition, the idea of agency is foregrounded; the suggestion is that, through critical thought about literature and the development of problem-solving abilities, students engaged in ecocritical thinking at school are empowered to make decisions about matters relating to the environment. The idea of responsibility is an important one. Garrard makes the point that for many people who regard themselves as environmentalists, there is an expectation of other authorities to address these concerns.

Regarding the possible choices of texts that are appropriate for teaching, one student stated,

> A literary text that can be used to teach this is Brenda Cooper's "For the [sake of] snake of power." This story is about a young woman who works at a public solar plant and lives in a low-income community that the company she works at serves. The story explores climate change and the impact that it has on low-income communities more than it has on rich communities. It also discusses how people can ensure that energy is contributed fairly. This can teach learners

> that even if they are wealthy, they should not ignore issues of global warming as it impacts poor people more.

This choice of text is of value because it addresses the topic of climate change and how human beings address the matter. Notably, this text falls into the genre of speculative fiction, unlike some of the other examples provided by students. Because it is part of the genre of speculative fiction, there is a sense that the text addresses matters in a manner in which the imagination is foregrounded. This type of text asks questions in the vein of "what if?" This type of questioning addresses fundamental concerns in our society.

In choosing to address indigenous poetry, another student gave this example:

> For example, if the learners are doing a poem that is based on the environment such as "An African Thunderstorm" which deals with aspects of nature, then questions may be raised such as the representation of nature in a specific stanza or the physical setting of the poem. In this case, The African thunderstorm represents the storm's impact on human existence and wind is a major theme in the poem.

In this case the student clearly links the lived experience (a thunderstorm) with a reader's interpretation. This poem refers to climate-related matters and the consequences of these for human beings. Thunderstorms are complex in nature—they bring water, which is life-sustaining, but they also bring elements of destruction, such as floods or gales. This multiple-edged element to weather is of value for exploration. Another aspect to consider is the relationship that humans have with such weather. The need for water is one such aspect. However, the destructive power of storms is another. Human engagements with storms include sheltering from a deluge, on a practical level. However, on an imaginative level there is the creation of mythological figures like Thor, the god of thunder. Out of the human relationship with weather such as thunderstorms the human imagination creates speculative and metaphorical responses that are understandable in their own terms—the mythological idiom.

In considering which texts to teach in the English class, one student made this observation:

> For example, literature such as *Tarzan of the Apes*, which was a series and part of 24 novels from 1912 to 1966 written by Edgar Rice Burroughs, portrays the story of a man who is divided between the human world and the natural world. As the story continues Tarzan the protagonist realizes the negative effect humans have on the environment and decides to side with animals because of how they respect him and the environment.

This student's response is of interest because Tarzan was not mentioned during the course of the module. Bearing in mind the well-known criticism that has developed over the years about the Tarzan books as texts that support ideas of racial inferiority (Tarzan, a white man, is the hero in the African jungle), it is interesting that a black South African student should make this choice. However, it is evident from the statement that the issue of race is not foregrounded in this commentary. Instead, the relationship that Tarzan has with the animals in the jungle is the main point of this claim. In this instance, Tarzan is a living example of the relationship between human and nonhuman living things, irrespective of his race. Notably, in the series of Tarzan books, many of the characters who are from the United States or Europe have an exploitative attitude to Africa; they wish to make a fortune out of the continent's natural resources. Because of this element in several of the books, it is possible to address the novels through the theory of cornucopia—the theory in which natural resources are regarded as "fair game."[39] The character Tarzan does not exploit the natural world. He has a greater connection to the natural environment in Africa than other people, because he was raised by an ape and is accepted as being one of the community. However, he is also descended from British aristocracy, and marries an American, so his life is a junction of different societies. It must be acknowledged that, although there are clear ecocritical elements in the Tarzan series of books, there is also an underlying concern about the representation of different race groups. Several of the depictions of black African people are problematic in these stories. However, there is an aspect of the ecocritical theoretical space that focuses on postcolonialism, and that could be an effective lens through which to scrutinize these texts.

In their commentary on South African ecocriticism Brooke Stanley and Walter Dana Philips address the themes that have tended to dominate this field of thought; there is a celebration of the landscape, and a recognition of the role of animals. These two writers point out that the Nobel Prize winner J. M. Coetzee has written about the landscape (and the issue of national identity and ownership of the land), as well as the issue of animals and how humans view them.[40] The two critics point out that "the fascination with megafauna" has worked to justify moving people off the land in order to create national parks.[41] Stanley and Philips make the point that in many cases the idea of ecocriticism is conceptualized in the United States or Europe, and then applied to South African literature and society. There is a need, they argue, for a South African ecocriticism, taking into account local circumstances. There is value in this approach, in that specific aspects of any given situation, including theoretical understandings of the environment, should inform our understanding of concepts. However, there is the danger of a parochial approach to a concept. In the same way that theoretical approaches that emphasize universalism tend to address topics in too broad a manner, by not taking into account specific elements that emerge in local situations, there is also the danger of ideas being too particular; they address matters in a way that is so specific that they are disengaged with the broader debate(s).

One of the students addressed Yann Martel's *Life of Pi*, which is a novel that is included in the Grade 12 curriculum as an option for study:

> A novel such as *Life of Pi* written by Yann Martel portrays a story of the protagonist Pi who because of various religious respects all living creatures on earth and only takes what he can to survive. Therefore, the *Life of Pi* novel can be used in English literature to teach sustainable utilization of natural resources.

This is an interesting observation. One of the readings provided as part of the presentation in the module was on Martel's novel; Elsie Cloete of the University of the Witwatersrand was the author of the text, and the focus was on the animal characterization in the novel and the notion of the other.[42] The novel is interesting from the point of view of genre—it includes a magical element (the carnivorous island), and other aspects of speculative fiction. The nature of the personal symbolism of the animals in the novel is of interest—Pi himself is the tiger (Richard Parker), if one accepts the alternative ending, and dual narrative, that is provided at the end of the novel, in which the animals each represent one of the human characters. There is a sense that the relationship that develops between Pi and the tiger is one in which the reader can accept that different species can coexist. This, however, is a complex matter. It is notable that the tiger walks into the jungle once the two shipwrecked survivors reach land, and Pi never sees him again. It could be the case that circumstances, rather than attitudes of inherent respect, resulted in the coexistence on the lifeboat.

This strong association of characters to animals (indeed, more of an identification than an association) makes this text an interesting one to study. It is notable that the animals in the novel are all part of a collection in a zoo; the role of zoos and the idea of animals as possessions are something to consider. Ownership of animals is regarded as acceptable these days, but many decades ago the ownership of humans by other humans was regarded as acceptable too. The issue of animals as possessions is therefore something to critique.

The topic of ecofeminism is of interest to several students. One student comments on ecofeminism thus:

> The eco-feminism theory states that women are viewed as people connected to nature and their emotions and men are viewed as cultural. The link ecofeminism shows between women link with nature can be used to teach the learners

in South Africa what will happen if the issues of gender-based violence go unanswered.

This student emphasizes the need to address the experiences and opinions of women, although this comment is not well-expressed. It is based on the idea that women are linked to nature, in that they are traditionally seen as nurturing, whereas men are linked to culture, in that they are associated with human intervention and invention; men are seen to bend nature to their purposes. There is also the sense that men are traditionally seen as rational, whereas women are instinctive and emotional. However, these binary understandings of the roles of people in societies are problematic, because they tend to present society as a fixed and closed entity. It is possible to argue that men are emotional beings, and that women are rational, despite the tendencies of socialization. It must be acknowledged, however, that there is little doubt that gender-based stereotypes still exist in societies across the planet.

The students did not limit themselves to texts intended for mature audiences, as the following example shows: "As the famous Dr Seuss said: 'unless someone like you cares a whole awful lot, nothing is going to change, it's not,' so why not teach the next generation to care 'a whole awful lot.'" This is a valid and appropriate summation of the commentary from literature relating to the environment. Two students addressed the Dr. Seuss text *The Lorax*, and this quote is from one of those responses. It is notable that this quotation is an exhortation to care about the world. It is also an exhortation to individual responsibility—there is the statement "someone like you," which points to the individual reading the text, and requires that person to take some form of responsibility for the state of the world. It is of interest that the motivating force for good is care rather than logic or rationality, although care could include an element of rationality. The concept of care includes a sense of moral judgment, and this connects with the previously discussed curriculum documents; those documents address social and environmental justice as something of value. Dr. Seuss, then, states that individuals must take responsibility for the state of the planet. This provides a sense of agency, and also demands a sense of personal commitment from the reader(s). At the very least, Dr. Seuss urges the reader to engage with ecocritical matters. This heightened awareness is a significant element of the field.

In the examples shared here the students engage with theoretical matters and approaches to teaching and learning in the classroom. The students explore literary texts and consider their value in the teaching of ecocriticism. Although their work is still evolving, there is evidence that the topics and themes of the module resonate with the students and their lived experiences.

CONSIDERING JOHN STEINBECK'S *THE GRAPES OF WRATH*

Steinbeck's *The Grapes of Wrath* is a well-known text that has bearing on ecocritical matters. The novel is not taught in this module, but I address it here because it is commonly taught in schools and can be considered from the position of providing a model for approaching some aspects of ecocritical theory. I will consider selected elements of the text in order to illustrate the ways in which readers could approach the novel through an ecocritical lens, specifically ecofeminism. The novel is vast in scope and covers many episodes in the lives of the Joad family, so there is a need to be selective when considering instructional time. A more comprehensive approach to the novel would require substantially more consideration. For this purpose I consider the first and last chapters, as the chapters that frame the events and themes of the novel, with a few minor references to other aspects.

In the first chapter of *The Grapes of Wrath* the reader encounters the relationship between nature and humans, as well as the relationships between groups of humans. The relationship between nature and humans is shown to be one of dependence, in that the humans—all of them farmers, or the families of farmers—are dependent on the cycles of nature to grow crops and thereby develop a sense of independence from other people, and thereby a sense of agency. In this situation nature is understood by humans in terms of human purposes. Humans plant crops in anticipation that nature provides rain, and that the rain will follow certain patterns of expectation. The change in the cycle of nature leads to a change in the rainfall, and this results in the beginnings of the dust bowl. Human purpose, which is dependent on the weather, is compromised. The crops on which the farmers depend are shown to grow, and suggest a promise of a future harvest, but then to wither and die as the anticipated rain fails to materialize.

The weather, and the change in the rain cycle, is dispassionate—it happens without awareness of human purpose, or any acknowledgment of the cataclysmic consequences of the lack of rain. Nature is seen to be removed from human experience or emotion. Human consciousness has constructed nature as a system of some predictability, and human behavior is shaped by this understanding of the systems. However, total understanding of weather and nature is not possible, and as a consequence humans act on beliefs almost as a matter of faith rather than knowledge.

The consequence of the dust bowl is that characters have to make decisions about how to act regarding their futures, as individuals and as communities. In this novel the human group is divided into categories relating to gender—in this case in terms of men and women—and also into categories of age—children and adults. In terms of the relationship of men and women, the former are presented by Steinbeck as decision-makers; it is the men who consider their options and make decisions. Indeed, it is these men who are the final focus of the chapter, in that they are shown to be "figurin'."[43] This intellectual or rational engagement will be the motive force that leads to a series of actions which culminates in the arrival of the Okies in California.

The women and children are less actively involved in the decision-making processes. They ask "What shall we do?"[44] but the answer is provided by the men. The women and the children look to the men for cues on how to behave.[45] Their behavior and state of mind is a reaction to the manner manifested by the men. Once they see that the men will not break, the previous status quo is restored; the children play and the women go about their domestic business.

The terms "thinking" and "figurin'" are significant. They appear in the final sentence of the first chapter.[46] These words are allusions to human rationality and suggest that, through thought, humans are able to solve problems. This problem-solving ability is presented as a male attribute. Women are presented as able to make judgments—they observe their men, to see if they will break. They do not, however, participate in the "figurin'," and are therefore not actively involved in ideas about how to deal with the calamity caused by the loss of their crops and the consequent loss of income.

These early pages of the novel present an environment overwhelmed by the dust—a consequence of the weather. This chapter is contrasted with the final sections of the novel, in which the Joad family and their peers are overwhelmed by rain. The rain holds echoes of the dust, in that the water seeps into the Joad home.[47] It is telling that in this situation there are no men "figurin'"—instead, Ma Joad takes charge of her family and urges them to move to higher ground. They take shelter in a barn.[48] It is evident that, at this point, Ma Joad is the decision-maker for the family, which is a shift from the first chapter. Also worthy of note is that the women in this chapter have names, unlike the nameless men in chapter one.

The final episode in the novel is that in which Rose of Sharon acts as a nurturer for an old, ill man. She feeds him the milk her body has produced (her baby is stillborn[49]) and she therefore provides him with life-giving sustenance.[50] Her action is almost instinctive—there is no need, no time for rational thought or "figurin'," she simply does what seems necessary.

In comparing the two chapters it is possible to explore a male versus female dichotomy. Men are presented as logical and rational, but they fail to find a solution for the challenges of the dust bowl. Their "figurin'" leads to their families traveling to California where they find themselves in difficult circumstances as migrant laborers. By contrast, the solution to the old man's dire state in the barn is found through instinct; there is no need for extensive figurin'. It is significant that the man who is associated with rationality, through his gender, is saved from death by the women who have been perceived as less-than-rational. By the end of the novel the women who were presented as observers in chapter one have become the initiators of action, and Rose of Sharon is presented as a savior.

This ecofeminist reading is admittedly selective—only two chapters have been addressed, but at the same time the elements of the chapters that have been considered are telling details. The way in which men and women respond to the challenges of the environment are clearly different, and the success of these responses to the challenges is also markedly different. Steinbeck seems to suggest that an instinctive response is perhaps better than rational thinking in addressing the calamities that humans face.

It is worth considering another element regarding the attitude toward nature and the differences in behavior presented by Steinbeck with regard to how men and women interact with it. The element is from chapter three, and it addresses the land turtle that attempts to cross the highway. The first car that comes down the highway is driven by a woman who sees the turtle and serves to avoid it. The second vehicle, a truck that comes down the highway, is driven by a young man, and he attempts to drive over the turtle. He is not successful, but clips the turtle's shell, throwing it off the highway by the rotational force of the tire.[51] The turtle then proceeds on its way.

Tom Joad picks up a turtle in a later chapter—there is a suggestion that it is the same turtle. In discussion with the preacher, Joad says that he intends to give the turtle to one of his younger brothers for a plaything.[52] It is telling that these men regard living things as toys. Again, the issue of human purpose is foregrounded. There is no consideration of the choices or purposes of the turtle itself (although it would be difficult to establish these). The turtle is regarded as part of the land's bounty to be collected by humans for their purposes only. From a deep ecology perspective the behavior and attitudes of these two men, and the driver of the truck, can be severely critiqued. Notably, the woman driver does not treat the turtle in the same way. She treats the turtle as a living thing that should be granted the opportunity to carry on its way.

Using deep ecological understandings of the relationship of humans with the environment, and with ecofeminism as a basis for critique, it is possible to explore Steinbeck's text in a manner that goes beyond the traditional close reading of literature such as that of the Cambridge school in the UK, associated with the work of F. R. Leavis, or the New Critics of the United States, associated with Cleanth Brooks, among others.[53] These two schools are well-established, but their foregrounding of formalism results in their disengagement from social or historical contexts. Ecofeminism provides insights that are not considered when using traditional approaches to literature. It is also possible to recognize some aspects of the cornucopian position in the novel—the land is regarded as valuable insofar as it provides an income or profit for those who live and work on it.

In approaching any text from an ecocritical perspective I suggest that the teacher/lecturer ensure that students have a basic knowledge of the theories that underpin the discipline. While not comprehensive, sharing knowledge of the basic aspects of the theory provides students with a degree of agency—they are able to approach the text and apply their knowledge themselves. The value of autonomous reading is well-established in education theory and practices.

In approaching this particular text I recommend focusing on the setting as a starting point. The setting addresses the landscape and the weather, which is a motive force in the novel itself. Beyond that, I suggest considering the social consequences of the setting. With the coming of the dust bowl the people of Oklahoma lose their income, their land, and their purpose. It is the decision to move to California that changes the roles of the Okies—they are no longer self-sufficient farmers, but instead migrant laborers, and this set of circumstances compromises their roles as human beings in society. They lose agency, self-respect, and hope, as they experience alienation. The change in the weather, depicted on the first three pages of the novel, has immense consequences for these human existences.

I also recommend a consideration of the attitudes that are expressed by the characters. The attitude to the turtle, for example, shows that not all human beings understand nature in the same way. Some treat it with respect, in that all living things are acknowledged, but others regard the turtle and other living things as resources. Considering the expression of such attitudes leads us to the question: How should nature be treated? What influences us in making these decisions? It is important to ask whose interests are being served. If we are to consider interests other than our own, there is a need for a sense of empathy. Questions that are asked about the characters and their decisions to act in particular ways can lead us to a sense of understanding and thus perhaps empathy. It is in developing answers to such questions that education processes take place. The questions, therefore, and the resultant discussion, are immensely significant.

CONCLUSION

It would be remiss of me not to address the words of one of the earliest ecocritical theorists—Rachel Carson. In *Silent Spring* she provides a stark warning:

> The most alarming of all man's assaults upon the environment is the contamination of air, earth, rivers, and sea with dangerous and even lethal materials.

> This pollution is for the most part irrecoverable; the chain of evil it initiates not only in the world that must support life but in living tissues is for the most part irreversible.[54]

Carson's recognition that human activity was having a disastrous effect on the environment holds a sense of urgency; if human beings do not change their behavior, inevitable negative outcomes loom.

In his memoir, David Attenborough emphasizes that the area of wilderness on the planet has been reducing steadily—from 66% of the planet's surface in 1937,[55] to 35% by 2020.[56] However, since the period in which Carson was writing there is greater acknowledgment of the need to address the environment in a manner that does not simply consider it a limitless supply of resources for human purposes. Attenborough points out that humans have faced many of the same challenges in the past. He stresses that for humans to succeed in addressing the challenges of the Anthropocene, we need to be united. He provides an example from 1986, in which nations that had a tradition of hunting whales decided to stop this practice to prevent the extinction of what he describes as "extraordinary and wonderful animals."[57]

Attenborough points to similar human interventions in central Africa, in which mountain gorilla populations have grown in size since nations agreed to establish sanctuaries.[58] There is an echo of this idea of national collaboration in Gaia Vince's commentary: "Into the Anthropocene countries will need to cooperate better on efficient energy generation, distribution, and storage."[59] The implication of Attenborough's and Vince's statements is that there is a possibility of success between humans (groups and individuals) if they engage in dialogue and discussion about these matters. There is a need to recognize this possibility, to address the matters and to debate, robustly, the values that inform our actions.

One voice that is very well-known in the field of climate change is the teenage activist Greta Thunberg. In commenting on the climate change challenges that humans face, she says, "I want you to act as if your house was on fire."[60] There is value in this request because it emphasizes the sense of urgency that should be adopted. However, it must be noted that panicking is not the best way to approach a problem-solving task. Instead of panicking, thoughtful collaboration is needed, as well as collective acknowledgment of the need for immediate action.

One approach to changing our understanding of the situation is to change our view of it. Nicholas Mirzoeff provides his readers with an interesting insight into how we see the world. He makes reference to the photograph of Earth taken by astronaut Jack Schmitt from Apollo 17 in 1972. The image is of the planet Earth, and is famously labeled *Blue Marble*.[61] The image provided humans with an unusual view of their planet, and made some think more carefully about the sphere on which they live. In 2012 NASA created another image of the earth, another Blue Marble. It differed from the former image because the twenty-first-century image was a composite view of the planet, made up of many images taken by satellites.[62] In both cases we are dealing with a human perspective of the planet. In both cases human technology was used to generate the image, but in the later depiction there is human agency in its creation, through positioning and coordinating many smaller images. This is an important point to highlight—we see the world in a particular way, and how we see it influences how we relate to it. Possibly we need to think of new ways in which we could see the planet. There is a comment ascribed to the science fiction author Arthur C. Clarke, in which he remarks, "How inappropriate to call this planet Earth when clearly it's Ocean."[63] His comment derives from the view that is offered from space, in which the oceans are seen to cover 71% of the planet's surface. When human perspectives change, it is possible to change our understanding of, or beliefs about, matters that we might tend to take for granted; the implication that human behavior could change under these conditions is what makes education such an important focus. It is how we change our perspectives.

There are elements of our environment that are in difficult circumstances, and it is clear that authors who address the human-environment relationship in their literary creations can help readers better understand these circumstances. These narrative accounts provide the reader with points of view, statements of events, a presentation of conflicts, and a possibility of resolution. To address these matters it is necessary for the reader to engage in debate, to question personal values, to question the moral codes of a group or an individual, and to make decisions based on what emerges. Literature provides insights that are a point of departure in the debates, but the final responses to the discussion could lie beyond the literature to related topics that connect to the reader's lived experience and world. Reading is therefore an end in itself, but also possibly a means to understanding the relationship between humans and the environment. In this way, literature becomes a motive force in the matter of human agency. Our actions are likely to be derived from and informed by our beliefs, and literary engagement provides us with an opportunity to explore and refine what we believe. Ecocritical theory and the application thereof in the school classroom therefore have a role to play in how students are to conduct themselves.

The Norwegian philosopher, Arne Naess, who is regarded as the father of deep ecology, proposed that the world's environment is diseased, but that, over a period of years, this situation can be remedied. However, the longer it takes to address the matter the longer it will take to heal.[64] Healing is only possible if humans change their behavior.

NOTES

1. Department of Basic Education, 2012. The Curriculum and Assessment Policy Statement (CAPS) Grades 10-12. Pretoria. Government Printer.
2. Department of Basic Education, 2008. The National Curriculum Statement (NCS) Grades 10-12. Pretoria. Government Printer.
3. Department of Basic Education, 13.
4. Department of Basic Education, 13.
5. Department of Basic Education, 12.
6. Department of Basic Education, 12.
7. Aryn Baker, "When the Taps Run Dry," *TIME Magazine* (February 19, 2018): 32.
8. Baker, "When the Taps Run Dry," 32.
9. Baker, "When the Taps Run Dry," 36.
10. Andre Kruger, *The Influence of Climate Change on the Value of Coastal Residential Properties in South Africa*. Unpublished Doctoral Thesis. University of Johannesburg, 2019: 27.
11. Kruger, *The Influence of Climate Change on the Value of Coastal Residential Properties in South Africa*, 36.
12. Kruger, *The Influence of Climate Change on the Value of Coastal Residential Properties in South Africa*, 39.
13. Kruger, *The Influence of Climate Change on the Value of Coastal Residential Properties in South Africa*, 1.
14. Kruger, *The Influence of Climate Change on the Value of Coastal Residential Properties in South Africa*, 1.
15. Kruger, *The Influence of Climate Change on the Value of Coastal Residential Properties in South Africa*, 41.
16. Bridget Martin, *Survival or Extinction: How to Save Elephants and Rhinos*. (Switzerland: Springer, 2019): 7.
17. Martin, *Survival or Extinction: How to Save Elephants and Rhinos*, 10.
18. Martin, *Survival or Extinction: How to Save Elephants and Rhinos*, 13.
19. Yolandi Groenewald, "The Exotic Game Market Goes Bang." *The Mail and Guardian* (January 11, 2019).
20. Groenewald, 2019.
21. Michael Bloomberg and Carl Pope, *Climate of Hope* (New York: St. Martin's Press, 2017), 90.
22. David Attenborough, *A Life on our Planet* (London: Witness books, 2020), 143.
23. Attenborough, *A Life on our Planet*, 143.
24. Cheryll Glotfelty, "Literary Studies in an Age of Environmental Crisis," in *Ecocriticism. The Essential Reader*, ed. Ken Hiltner (Routledge: London, 2015), 122.
25. Timothy Clark, *The Cambridge Introduction to Literature and the Environment* (Cambridge University Press, 2011): 4.
26. Greg Garrard, *Ecocriticism* (London: Routledge, 2012).
27. Garrard, *Ecocriticism*, 21.
28. Garrard, *Ecocriticism*, 21.

29. Garrard, *Ecocriticism*, 23.

30. Garrard, *Ecocriticism*, 26.

31. J. M Coetzee, *The Life and Times of Michael K.* (Johannesburg: Ravan Press, 1983).

32. Coetzee, *The Life and Times of Michael K.*, 5.

33. Coetzee, *The Life and Times of Michael K.*, 1.

34. Coetzee, *The Life and Times of Michael K.*, 250.

35. Coetzee, *The Lives of Animals* (Princeton: Princeton University Press, 1999).

36. Jenny Robson, *Savanah 2116* (Cape Town: Tafelberg, 2004).

37. Note that student responses have been included verbatim.

38. Alan Paton, *Cry, the Beloved Country* (London: Jonathan Cape, 1948).

39. Garrard, *Ecocriticism.*

40. Brooke Stanley and Walter Dana Philips, "South African Ecocriticism: Landscapes, Animals and Environmental Justice," *Oxford Handbooks* online (2017): 4.

41. Stanley and Philips, "South African Ecocriticism: Landscapes, Animals and Environmental Justice," 6.

42. Elsie Cloete, "Tigers, Humans, and Animots," *Journal of Literary Studies* 23, no. 3 (2007): 314–333.

43. John Steinbeck, *The Grapes of Wrath* (London: Penguin, 1939): 6.

44. Steinbeck, *The Grapes of Wrath*, 6.

45. Steinbeck, *The Grapes of Wrath*, 6.

46. Steinbeck, *The Grapes of Wrath*, 6.

47. Steinbeck, *The Grapes of Wrath*, 530.

48. Steinbeck, *The Grapes of Wrath*, 533.

49. Steinbeck, *The Grapes of Wrath*, 523.

50. Steinbeck, *The Grapes of Wrath*, 536.

51. Steinbeck, *The Grapes of Wrath*, 18.

52. Steinbeck, *The Grapes of Wrath*, 24.

53. Raman Selden and Peter Widdowson, *A Reader's Guide to Contemporary Literary Theory* (Lexington: The University Press of Kentucky, 1993): 15–16.

54. Rachel Carson, *Silent Spring* (New York: Mariner Books, 1961): 6.

55. Attenborough, *A Life on Our Planet* (London: Witness, 2020): 13.

56. Attenborough, *A Life on Our Planet*, 95.

57. Attenborough, *A Life on Our Planet*, 216.

58. Attenborough, *A Life on Our Planet*, 217.

59. Gaia Vince, *Adventures in the Anthropocene* (London: Vintage, 2014): 345.

60. Greta Thunberg, *No One is Too Small to Make a Difference* (London: Penguin, 2019): 45.

61. Nicholas Mirzoeff, *How to See the World* (London: Penguin, 2015): 3.

62. Mirzoeff, *How to See the World*, 9.

63. Alison George, ed. *This is Planet Earth: Your Ultimate Guide to the Place We Call Home, New Scientist Instant Expert* (John Murray Learning, 2018): 149.

64. Arne Naess, *Ecology of Wisdom* (London: Penguin, 2008): 309.

REFERENCES

Attenborough, David. *A Life on our Planet*. London: Witness books, 2020.

Baker, Aryn. "When the Taps Run Dry." *TIME Magazine*. February 19, 2018.

Bloomberg, Michael and Pope, Carl. *Climate of Hope*. New York: St Martin's Press, 2017.

Carson, Rachel. *Silent Spring*. New York: Mariner Books, 1961.

Clark, Timothy. *The Cambridge Introduction to Literature and the Environment*. Cambridge: Cambridge University Press, 2011.

Cloete, Elsie. "Tigers, Humans, and Animots." *Journal of Literary Studies* 23, no 3. (2007): 314–333.

Coetzee, J. M. *The Life and Times of Michael K*. Johannesburg: Ravan Press, 1983.

Coetzee, J. M. *The Lives of Animals*. Princeton: Princeton University Press, 1999.

Department of Basic Education. *The National Curriculum Statement (NCS) Grades 10–12*. Pretoria. Government Printer, 2008.

Department of Basic Education. *The Curriculum and Assessment Policy Statement (CAPS) Grades 10–12*. Pretoria. Government Printer, 2012.

Garrard, Greg. *Ecocriticism*. 2nd ed. London: Routledge, 2012.

George, Alison, ed. *This is Planet Earth: Your Ultimate Guide to the Word We Call Home. New Scientist Instant Expert*. London: John Murray Learning, 2018.

Glotfelty, Cheryll. "Literary Studies in an Age of Environmental Crisis." In *Ecocriticism: The Essential Reader*, 120–129. Edited by Ken Hiltner. Routledge: London, 2015.

Groenewald, Yolandi. "The Exotic Game Market Goes Bang." *The Mail and Guardian*, January 11, 2019.

Kruger, Andre. *The Influence of Climate Change on the Value of Coastal Residential Properties in South Africa*. Unpublished Doctoral Thesis. University of Johannesburg, 2019.

Martin, Bridget. *Survival or Extinction: How to Save Elephants and Rhinos*. Switzerland: Springer, 2019.

Mirzoeff, Nicholas. *How to See the World*. London: Penguin, 2015.

Naess, Arne. *Ecology of Wisdom*. London: Penguin, 2008.

Paton, Alan. *Cry, the Beloved Country*. London: Jonathan Cape, 1948.

Robson, Jenny. *Savanah 2116*. Cape Town: Tafelberg, 2004.

Selden, Raman and Peter Widdowson. *A Reader's Guide to Contemporary Literary Theory*, 3rd ed. Lexington: The University of Kentucky Press, 1993.

Stanley, Brooke and Walter Dana Philips. *South African Ecocriticism: Landscapes, Animals and Environmental Justice*. Oxford Handbooks online, 2017.

Steinbeck, John. *The Grapes of Wrath*. London: Penguin, 1939.

Thunberg, Greta. *No One is Too Small to Make a Difference*. London: Penguin, 2019.

Vince, Gaia. *Adventures in the Anthropocene*. London: Vintage, 2014.

Chapter 6

These Are the Forgeries of Jealousy

Nature Out of Balance

Timothy J. Duggan and Natalie Valentín-Espiet

And this same progeny of euills,
Comes from our debate, from our dissention,
We are their parents and originall.[1]

—Titania, *A Midsummer Night's Dream*

This chapter seeks to explain how students may examine the implications of human behavior and, particularly, inharmonious relationships as causal agents of negative climate change through study and embodied representation of a monologue from Shakespeare's *A Midsummer Night's Dream* (*MND*).[2] The chapter also will demonstrate the efficacy of embodied learning, engaging theater techniques to elicit student comprehension, textual connections, and a range of student work products, from written scripts to original performances. Such inquiry will allow students to investigate collaboratively a number of environmental concerns, from disruptions in cultural events during inclement weather to outright environmental disasters. Specifically, this layered workshop is designed to explore the intersection between Titania's "These Are the Forgeries of Jealousy" speech in Act 2, scene 1 of *MND* (81–117)[3] and contemporary manifestations of climate change. Through combined group analysis of Shakespeare's text and physical interpretation using theater strategies, students may bring to life small chunks of the speech in a way that helps them understand not only Titania's message to Oberon in the play, but Shakespeare's possible message about human agency in natural disaster. Furthermore, the activities described in this chapter will connect the individual beats in the monologue with contemporary informational texts describing current climate problems. Students will study those informational texts collaboratively and incorporate them into their performance of Titania's speech.

With the publication in 2017 of their book, *Teaching Climate Change to Adolescents*, Beach, Share, and Webb urged all English teachers to address environmental issues in our curriculum, arguing that the scientific community has well established the existence and particular human-generated causes of climate change.[4] Therefore, they argue, the necessary work of shifting public attitudes and instilling a sense of urgency to address causes of potential environmental collapse falls to the agency of English teachers, whose particular domain of expertise is literacy and communication.[5] With the concurrent publication of the U.S. Government's "Climate Science Special Report" detailing indicators of human responsibility for climate change as manifested by extreme weather events, melting sea ice, ocean acidification, and other potentially disastrous problems,[6] teachers have compelling evidence to justify designing curriculum that allows students to learn not just the science, but the political and social aspects of a changing climate.

English teachers are not without scholarly frameworks when attempting to address climate change in their classes. While the connections between humans and nature as represented in literature, most notably in the Romantic Era, have long existed, interest in nature writing of the past century, particularly Rachel Carson's groundbreaking work, *Silent Spring* (1962), and the works of literary nonfiction writers such as Annie Dillard, Barry Lopez, and Kenneth Brower engendered critical responses from the scholarly community, creating an intersection of science writers and literary critics against the backdrop of the environmental movement. These scholars eventually organized and formed the Association for the Study of Literature and the Environment (ASLE) in 1992.[7] The group hosts a biennial conference, convenes various symposia, and produces a peer-reviewed academic journal that features works in ecocriticism as well as original creative works. According to their website, members of ASLE "study, write, compose and create because [they] care about issues like biodiversity, environmental justice, survival in a time of endemic precarity and global catastrophe, and the effects of climate change on humans and nonhumans alike."[8] The field of ecocriticism has expanded over the past three decades, in part due to the growing recognizability of the climate crisis, and in part because of the growth of interest in scholarly study of nature literature, climate justice, and interdependence.

Shakespeare, being situated squarely in the center of the traditional English canon, and being ubiquitous in contemporary school curricula and theater culture, has not escaped the notice of ecocritical scholars. In 2006, Gabriel Egan's book, *Green Shakespeare*, made a case for new approaches to understanding Shakespeare's complex relationship to nature.[9] In explaining the rationale for writing his book, Egan referenced the 2005 meeting of the Shakespeare Association of America (SAA) in Bermuda. He mentioned that year as a watershed in public awareness and concern over climate change.[10]

And yet, Egan complained, not one session at the SAA conference had any connection to the environment.[11] His purpose in writing the book, therefore, was to bring ecocriticism to the study of Shakespeare,[12] and so he did. For the past several years, panels and seminars involving ecocriticism are an annual feature of the SAA conference.[13]

One tenet of Shakespeare ecocriticism is to avoid, whenever possible, superimposing current environmental ethics onto Shakespeare's 400-year-old texts, or worse, attempting to rationalize Shakespeare himself as an "environmentalist" in the modern sense of the term. This is not to say that studying Shakespeare through an ecocritical lens is not useful in the present. Shakespeare, after all, had a deep understanding of nature, and the imagination to invest particulars in nature with human characteristics. As Simon Estok points out, "Applied to Shakespeare, [ecocriticism] extends huge amounts of foundational work that has already been done into exciting new areas. It can help us to understand where we have come from, where we are, and where we might be going."[14] Compelling cases have been made for Shakespeare's connection with nature as understood by his contemporaries, such as Ben Jonson and Robert Greene, and his anthologizers, Heminges and Condell.[15] Elizabethan understanding of humanity's place in the natural order was most influentially described in *The Elizabethan World Picture* by E. M. W. Tillyard.[16] Tillyard's reconstruction of the natural order includes a tightly balanced network of correspondences between nature as understood at the dawn of modern scientific thinking and human society in the age of "divinely" ordained queens and kings.[17] The Elizabethans learned the bodily humors identified by Hippocrates and Galen (blood, yellow bile, black bile, and phlegm) as they theoretically corresponded with the natural elements of air, fire, earth, and water. As Tillyard presents it, the correspondence was part of a "Chain of Being" that reflected a harmonious world expressing "the unimaginable plenitude of God's creation, its unfaltering order, and its ultimate unity."[18] And yet, in examining the descriptions of Shakespeare by those who followed him, most notably Milton and Samuel Johnson, who coined the term "Sweet Swan of Avon," Steve Mentz claims that beneath these tributes lies "a naïve attitude toward a nature that is assumed to be beneficent, harmonious, and green."[19] Mentz confronts portrayals of nature in Shakespeare that are anything but harmonious. Certainly, the blasted heath and equivocating fiends in *Macbeth*, the ominous and unnatural storms in *King Lear* and *Julius Caesar*, the disruption of natural order in *Troilus and Cressida* have each provided material for ecocriticism, as have the utterances of particular characters in Shakespeare, such as Jaques, the melancholic, in *As You Like It*.[20]

It is one monologue delivered by Titania, Queen of the Faeries, in *MND*, during an argument with her estranged husband, Oberon, King of the Faeries, that is the focus of this chapter's workshop. Just prior to the speech, Titania has accused Oberon of intruding on her company because of his love for Hippolyta, Queen of the Amazons, who is to marry Duke Theseus of Athens. She intimates a sexual affair between Oberon and Hippolyta, and Oberon counters by accusing Titania of having a sexual relationship with Theseus. Titania responds to Oberon's accusation with this speech:

TITANIA
These are the forgeries of jealousy,
And never, since the middle summer's spring
Met we on hill, in dale, forest or mead,
By paved fountain or by rushy brook,
Or in the beached margent of the sea
To dance our ringlets to the whistling wind . . .

We are their parents and original.[21]

Titania's monologue is typical of Shakespeare's writing style in the mid-1590s, the period during which most scholars place the composition and first performance of *MND*. Long speeches with image-laden, often romantic expressions flavor much of the play and its compositional twin, *Romeo and Juliet*. In performance, the monologue has typically gone the way of other long speeches in Shakespeare that do not drive action—it has been cut, or at least partially cut, from the script. In the famous 1935 film that featured Mickey Rooney as Puck and a beautiful soundtrack by Mendelsohn, the speech is cut entirely.[22] In the 1996 film of a performance by the Royal Shakespeare Company, roughly half of it is cut,[23] and in the 1999 film featuring Michelle Pfeiffer as Titania, only the first twelve lines and the final three lines are kept,[24] with no cinematic gesture toward the imagery of the speech—only thunderclaps. Likewise, teaching guides often focus attention on the lovers in the play, the mechanicals, and the magical aspects of the Faerie Kingdom. The popular *Shakespeare Set Free* series authored by associates of the Folger Shakespeare Library education program includes a volume devoted to *MND*, *Romeo and Juliet*, and *Macbeth*.[25] The lesson in the book devoted to Act 2, scene 1 does not mention the catalogue of environmental problems articulated by Titania.[26] In his *ShakesFear and How to Cure It*, Ralph Cohen dismisses the significance of this monologue, advising, "You may wish to condense Titania's speech about the weather."[27]

This seeming lack of attention to the monologue, however, is not universal. In the RSC's 1968 film starring Judy Dench as Titania, all the lines are kept.[28] In Julie Taymor's 2014 movie version of her stage play, the speech (all but three lines of it) remains,[29] with gloomy music and a sweeping backdrop of green and orange clouds. Close-up camera shots of Oberon indicate facial recognition of his complicity in the natural imbalance. Casey Wilder

Mott's campy 2018 reframing of the story in Los Angeles[30] goes a step further in connecting Titania's monologue to climate change. Most of the lines—though not all—are retained, and Mott infuses actress Mia Doi Todd's delivery with camera images of environmental catastrophe, such as hurricanes, factory-fueled air pollution, extreme drought, flooding, crop failure, and upheaval of seasons.

Mott's cinematic gesture toward the climate change theme is not without scholarly antecedent. Regina Buccola's 2010 critical guide to the play includes an essay by Tripthi Pillai that addresses Titania's monologue directly:

> In a speech conjuring images that eerily anticipate twenty-first century horrors of global climate change, the queen of fairies claims joint responsibility with Oberon for the various sea storms and floods that have ruined the crop and rendered useless the agrarian economy that depends on the systematic notion of seasons, which are the climatic representation of the passage of time.[31]

Pillai goes on to argue that Titania's assertion of herself and Oberon as causal agents in the climatic upheaval that so disrupts human life demonstrates that it is "folly" for the human characters in the play to believe that they control their own futures.[32] Another way to see this speech is to consider the anthropomorphization of Titania and Oberon and their interpersonal disharmony as an example of human agency in climatic disruption. A similar point is made by Sophie Chiari, who argues that Titania, "who always sounds very much like a human," ascribes the "poor climate and natural catastrophes" to the discord in her relationship with Oberon.[33] As Chiari points out, while the norm in Shakespearean representations of nature out of balance is to cast human characters as victims, this speech shows an anthropomorphic deity claiming responsibility for a host of climate plagues (so lengthy that even Chiari excerpts them in her analysis).[34] It is this long catalogue of maladies in Titania's monologue that this chapter's workshop explores, first using performance activities to create student engagement with Shakespeare's text, and then using textual connections to contemporary informational texts on climate change to develop student consciousness of current climate disasters and the human agency that engendered them.

But why use performance strategies as opposed to traditional literary analysis to make this connection? The performance approach to studying Shakespeare is well-established. The integration of theater techniques in English classes dates back to the 1960s, when Homer Swander "electrified and surprised the annual meeting of the Shakespeare Association of America with his experiments in teaching Shakespeare through performance."[35] According to Swander, "To read a script well is to discover what at each moment it tells the actor or actors to do."[36] Edward Rocklin, in his *Performance Approaches to Teaching Shakespeare*, went a step further in declaring,

> I think we can get at . . . meaning more fully if we *start* by asking, "What do these words do?" and then, "What can these words be made to do" and then, "What do these words make the actors make the audience do?"[37] Emphasis in original

The publication by MLA in 1999 of *Teaching Shakespeare through Performance*, edited by Milla Cozart Riggio, brought a wide variety of voices to the argument for incorporating drama strategies to teach Shakespeare in English class.[38] In the decades since, many partnerships have flourished between theater companies and schools through artist residencies and workshops, similar to those offered by the education program of the Chicago Shakespeare Theater, where this particular workshop originated. Likewise, the Folger Shakespeare Library's education wing, under the direction of Peggy O'Brien, emphasizes the connection between performance exercises and literary learning. The Folger has published their *Shakespeare Set Free* series of books to encourage such activities.[39] According to Duggan, embodying Shakespeare's texts through drama workshops in the English classroom "changes students' relationship with the literary work, with the teacher, and with each other."[40] Doing so also diversifies modalities through which students experience text, creating a synergy of text, vocalization, and movement.

Furthermore, using drama to explore climate change in English class is supported by the work of Beach, Share, and Webb, as they encourage teachers to integrate role-play and other drama strategies to engage students in conversations about a changing planet, modeled on the work of Augusto Boal in his *Theater of the Oppressed*.[41] In the workshop presented below, theater activities help students to understand the referent Shakespeare text as well as the matched informational texts on climate change.

THE WORKSHOP

Part One: Shakespeare's Text

The activities described here may be nested within an instructional unit on climate change or addressed within a thematic unit with *MND* as a core text in a survey class. The amount of class time required for the workshop will vary, depending upon the kinds of follow-up and assessment teachers choose. The lesson plan accompanying this chapter spans three days of instruction.

To begin, the teacher should pull Titania's speech in its entirety from the script of *MND* so that students have it in their hands while working on their feet.[42] A first read-through may be done silently or as a read-aloud, to get a sense of the structure and sounds in the monologue. In the first seven lines, Titania describes how Oberon has disrupted the "ringlets" or dances with his "brawls," a violent action which Titania identifies as the cause of the climatic upheaval. The pivotal word "Therefore" shifts attention from cause to effect, and it is here that the remainder of the speech may be broken into small chunks or "beats," starting with, "Therefore the winds, piping to us in vain," (2.1.88) and ending with, "We are their parents and original" (2.1.117). Titania's list of natural disasters divides nicely into twelve beats, each identifying a different climate problem. In the lesson plan that accompanies this chapter, the entire speech is printed with boundaries between beats indicated by alternating regular and bold-faced print. Footnote markers after each identify the climate problem mentioned by Titania.

Having divided the monologue into the small chunks or "beats," the teacher should assign each beat, starting with line 88, to groups of three to four students. To initiate the feel of a workshop in class, teachers may first want to set up a tableau vivant of Titania and Oberon, with two students to play the roles, and then have the student representing Titania read the speech in its entirety. A good next step is to have students stand in a circle around the tableau figures as the class reads the lines together, with the tableau Titania reading the first seven lines, followed by each group reading their assigned section. At this point, the teacher should tell students that they will be responsible for performing their beat by physically representing the natural disaster described therein.[43] Shakespeare's use of personification in describing the different climate problems will help students to embody the text and create an interpretation. Here are some suggestions for coaching students in their physical interpretations:

Therefore the winds, piping to us in vain,
As in revenge have sucked up from the sea
Contagious fogs[44]

Suggestion: The winds are personified as pipers, "piping in vain," ostensibly because humans are not listening. Therefore, the winds convert their pipes into straws to "suck up from the sea / contagious fogs" and spread them across the land. Students can embody these actions.

which, falling in the land,
Have every pelting river made so proud
That they have overborne their continents.[45]

Suggestion: Notice that the rivers are assigned the human emotion of pride—pride that causes the rivers to break the confines of their banks. Students can physicalize that pride and the resulting flood in a number of ways.

The ox hath therefore stretched his yoke in vain,
The ploughman lost his sweat, and the green corn
Hath rotted ere his youth attained a beard.[46]

Suggestion: There are three figures in this beat: the ox, the ploughman, and the personified corn. Students may interpret the actions assigned to each figure or otherwise represent crop failure.

The fold stands empty in the drowned field,
And crows are fatted with the murrain flock.[47]

Suggestion: A fold is a pen or enclosure for livestock, typically sheep. A "murrain flock" is a group of animals that have been killed by disease. In this case, the dead animals are eaten by crows.

The nine men's morris is filled up with mud,
And the quaint mazes in the wanton green
For lack of tread are undistinguishable.[48]

Suggestion: This is the least consequential of the environmental problems cataloged in this speech, as it refers to the disruption of outdoor recreational activities due to persistent inclement weather. Nine men's morris is an ancient board game that, in England, was sometimes played on an outdoor field. The English were also fond of natural mazes of hedgework. A contemporary equivalent to this environmental issue would be the cancellation of professional sporting events due to hurricanes and other natural disasters. Students may want to embody sports with which they are more familiar, rather than what appears in the speech.

The human mortals want their winter cheer.[49]
No night is now with hymn or carol blest.[50]

Suggestion: Students could portray carolers or people engaging in happy winter activities and then having them disrupted. The beat describes human misery, so there are many options for students to physicalize that misery.

Therefore the moon, the governess of floods,
Pale in her anger washes all the air,
That rheumatic diseases do abound;[51]

Suggestion: This section refers to flood-borne diseases, such as cholera and dysentery. The moon is personified as an angry female figure, so some kind of angry motion with accompanying representations of illness would work well.

And thorough this distemperature we see
The seasons alter: hoary-headed frosts
Fall in the fresh lap of the crimson rose,[52]

Suggestion: Any movement or positioning that indicates an upheaval of seasons would work. Unseasonable weather can occur in winter or in summer. Students can use their imaginations to embody the frost falling in the "lap" of a rose.

And on old Hiems' thin and icy crown
An odorous chaplet of sweet summer buds
Is, as in mock'ry, set.[53]

Suggestion: Hiems is a Latin word for winter, which here is anthropomorphized as an old king, perhaps "Old Man Winter." One student could represent the king and others could strew real or imaginary flowers while making mocking gestures.

The spring, the summer,
The childing autumn, angry winter, change
Their wonted liveries,[54]

Suggestion: This beat also refers to the seasons altering, through the metaphor of the seasons changing clothes with one another.

and the mazed world,
By their increase now knows not which is which;[55]

Suggestion: Think of the word "mazed" as "amazed" to make sense of this passage. Confusion caused by climate events can lead to more accidents and misery. The pronoun "their" refers to the previous passage personifying the seasons. Another interesting word here is "increase," which may refer to intensifications of seasonal weather, such as extreme heat and cold, more intense storms, and so on. The "increase" may also refer to crop yield.

And this same progeny of evils comes
From our debate, from our dissension.[56]

Suggestion: Here is where Titania places the blame for all of the cataloged environmental problems on Oberon and herself, claiming that their "debate" and "dissension" are the cause. If we think of Titania as having human form and characteristics, we can link this beat to the claim of human-caused climate change. Students may want to embody the "progeny of evils," something they will enjoy doing, or enact a quarrel.

We are their parents and original.[57]

Suggestion: This line might be read by the entire class together in unison, as a fit closure to the speech. The class may even come up with a shared final gesture to sum up their portrayal, maybe all falling down to indicate death or joining hands in a gesture of solidarity. Other performance options for the last line can be generated by students.

As mentioned above, these enactments would best be done with the class standing in a circle, and individual groups standing together to play their parts in succession while reciting the words. Variations include having each group quickly move to the center of the circle when their turn comes. In that case, they would not need to be standing together while other groups take their turns. For best results, teachers can take their students to an outdoor space for the choral performance of the speech. In a virtual environment, breakout rooms would allow students to devise ways to use the visible space of their screens to create the illusion of these climate manifestations. Because each group has only a couple of lines to interpret physically, they can brainstorm their movements and characterizations fairly quickly, say in ten to fifteen minutes. They may also want to rehearse their parts and memorize the lines, depending on how much time the teacher allows them. The important work is in their rehearsal and decision-making in acting out the speech. The first time through with the entire class may be clunky, so the teacher should be prepared to have students perform the speech with the group embodiments more than once.

If the workshop were to end here, the class could discuss what they've learned about the speech, Shakespeare's imagery, and their performance choices. Stopping at this point, however, would deprive students of the opportunity to inquire further into the specific disasters cataloged in the speech. In order to shift the focus from the monologue, and specifically, to allow students to dive more deeply into the issues the speech raises about climate change, the activities described below will provide strategies for further learning.

Part Two: Informational Texts

All teachers, including English teachers, have a mandate to incorporate informational texts into their instruction. In the United States, Common Core

Standards for English Language Arts privilege informational texts in their recommended reading regimen,[58] and most English teachers recognize the importance of connecting students with scholarly interpretations of literature found in professional literary journals. While we rightfully often think of informational texts related to the environment as scientific literature, many of the more dramatic and unfortunate problems related to a changing climate make it into local and national news, public service announcements, and medical information sources. The term "climate change" is increasingly part of the public vernacular, which in itself is a sign that climate change is moving forward in the public consciousness. When students perform climate problems, as they have in embodying Titania's speech, the activity creates possibilities for further study of the climate problems as they exist now, rather than in Shakespeare's day.

One problem that students face when trying to learn about climate change is that the issue is too broad and too complicated to wrap one's mind around with clarity. Therefore, an advantage of this workshop approach using Titania's monologue is that, by breaking the speech into its constituent parts and then assigning those parts to small groups of students for further investigation, we narrow each group's focus onto one problem related to climate change, giving each group a separate and specific line of inquiry. The amount of responsibility or independence teachers afford their students in pursuing identification and investigation of the natural disasters described in the monologue requires a judgment call on the part of the teacher, based on students' prior knowledge of climate change and their research skills. As a scaffold, the speech printed in the lesson plan that accompanies this chapter lists the climate problems mentioned by Titania, and teachers are welcome to use that list. Furthermore, links to short (one-page) informational texts associated with those problems appear in the lesson plan. To save time, a teacher may wish to have students access those texts and proceed to the next step in the workshop. But teachers who want to give students more authority in the process may choose not to include the identified list of climate problems provided in the lesson plan and, instead, have their students identify the various manifestations of climate change Titania describes on their own. In that scenario, students would then research the climate problems they have identified and bring to class materials they have accessed to their small groups for further study. Once the list of climate problems addressed is assembled and students have accessed current informational texts related to those problems (or once they have been given informational texts by the teacher), they may read, annotate, discuss, and prepare to share their learning through augmented performance.

Students will collaborate with their group members to comprehend the informational texts and to explore the connection between the informational material and Titania's monologue. To prepare students for the next step, teachers may tell them that during the next read-through of the speech, they are going to either summarize their informational text into a written script or pick selected quotes from it to read and perform. Students will need time to develop their scripts and performance choices. For example, one group of students will have the section of the monologue below:

Therefore the winds, piping to us in vain,
As in revenge have sucked up from the sea
Contagious fogs[59]

This beat refers to the winds drawing "contagious fogs" from the sea. The description is suggestive of toxic air pollution. The American Lung Association (ALA) has issued the following statement on toxic air pollution, articulating its connection to human industrial practices.

What Are the Health Effects from Toxic Air Pollutants?

Toxic air pollutants pose different risks to health depending on the specific pollutant, including:

- Cancer, including lung, kidney, bone, stomach
- Harm to the nervous system and brain
- Birth defects
- Irritation to the eyes, nose and throat
- Coughing and wheezing
- Impaired lung function
- Harm to the cardiovascular system
- Reduced fertility

How Are People Exposed to These Pollutants?

> People inhale many of these pollutants in the air where they live. But, since these pollutants also settle into waterways, streams, rivers and lakes, people can drink them in the water or eat them in the fish from these waters. Some hazardous pollutants settle into the dirt that children play in and may put in their mouths.

Where Do Toxic Air Pollutants Come From?

> Major sources of toxic air pollutants outdoors include emissions from coal-fired power plants, industries, and refineries, as well as from cars, trucks and buses.[60]

Using the information from the ALA as the source, a group of students may generate a summary statement as a script for performance:

> Toxic air pollution causes health problems like cancer, harm to the nervous system, birth defects, irritation to the eyes, nose, and throat, coughing and wheezing, impaired lung function, harm to the cardiovascular system, and reduced fertility. People inhale these pollutants in the air. Pollutants come from coal-fired power plants, industries, and refineries, and motor vehicles.

Once their script is written, the group may read it in choral fashion while acting out the health problems or the sources of pollution mentioned in their informational text. They may also choose to have one member of the group read the script while remaining members act out what is being read. Students may instead develop a devised scene, such as a TV announcer reading the summary, or they may write a dialogue between members of the group discussing the content of the informational text. They may also excerpt lines or phrases from the informational text to perform. Teachers who give students time to develop these scenes will have plenty of material to discuss and will increase student understanding of the informational texts from which their scenes are devised.

When students are ready, the class as a group can perform Titania's monologue again, this time with the addition of their informational scripts. As each group recites their section of Titania's speech, they will also perform their informational text script before the next group performs their section. It may be wise to have a universal "pass it on" signal so that each succeeding group knows when the group before them is finished with their segment. Once these performances are completed, and students have experienced the connection between the original *MND* speech and the informational texts, the class may discuss what they have learned before proceeding to the final performance-related step in the workshop.

The final iteration of the group performance in this workshop demonstrates how to incorporate the knowledge gained through the informational texts into a dramatic reading of Titania's monologue without the long delays required to accommodate the performance of the informational texts. A clear way to do so is through distillation of the informational material into single key words from those texts that students may add to Titania's speech. Teachers may have students pick only one to five keywords from their constructed speeches/performances of the informational text, so that the flow of Titania's monologue is not greatly interrupted. These key words become "echoes" of Titania with words pulled from the informational texts.[61] Students may perform a few of the same physical actions as they performed in the first choral reading, but also vocalize these selected "echoes" from the informational texts after they read Titania's words.

Using the example informational text from the ALA above, students may choose the following words (as an example, not a rule): cancer, irritation, harm, lungs, baby. When they perform the monologue, their script would run as follows:

Therefore the winds, piping to us in vain,
As in revenge have sucked up from the sea
Contagious fogs[62] (Echo: cancer, irritation, harm, lungs, baby)

Students may add sound effects indicated by the informational text, such as coughing and, perhaps, expressions of grief. Performing the script this way will not greatly impede the progression of Titania's speech and may have a powerful emotional effect as each group of students performs their lines and echoes. When this step in the workshop is completed, the class may again discuss what they have learned, and how pairing the speech from the play and the informational texts furthered their understanding of both texts.

Following the workshop, students may write reflectively about what perspectives they have gained on climate change, how the drama workshop engaged them in the texts and the issues portrayed, and how working collaboratively helped them to understand the texts and their implications. Assessments for these activities may lie in the performances themselves or in the reflective writing, depending on the larger context within which this workshop is conducted. The lesson plan that accompanies this chapter includes other prompts for discussion and writing.

As described here and in the lesson plan, the workshop would require roughly two to three days of instruction in a standard, fifty-minute daily class schedule. Being mindful of the vicissitudes of available space in the curriculum for deep dives into material as balanced by other curricular priorities, within the structure of this workshop lie openings for cutting or condensing steps in order to complete the lesson more quickly. Teachers may decide to skip the full performance of the informational texts and go directly from the first choral performance of the scene to the "echoes" performance. Furthermore, teachers will cut down the duration of the workshop by providing the suggested short informational texts corresponding to the beats in the monologue, thus eliminating the time students would spend researching their climate problem on their own.[63] One caveat of condensing the workshop would be loss of student engagement and analysis of the informational texts themselves, which lie at the heart of the workshop. After all, if the primary goal of the workshop is to use Shakespeare as a vehicle for investigation of current climate issues (and not the other way around), the time spent with the informational texts must be preserved. Consequently, if more time is available, students may engage in further research into the particular forms of climate-related problems expressed in the monologue. Students may also discuss or write about what they have learned about themselves from this deep dive into text. They may comment on how physicalizing the informational text and linking it to the speech made the information more "real" for them, or they may want to discuss the concept of human agency in climate change and write about it reflectively. Another approach to post-activity discussion is to ask students what contemporary climate issues are *not* included in Titania's speech, such as wildfires, intense heat, melting of the polar ice caps, and so on. Students could rewrite the speech as though it were given today, and such a writing or performance assignment could lead to further inquiry into current manifestations of climate change.

CONCLUSION

The workshop described in this chapter shows only one possibility for marrying literary texts with informational texts to build student comprehension and awareness of climate change. The collaborative and physical or "embodied" learning that this workshop requires gets students out of their seats and puts them in the shoes of actors. But beyond that, students also get to make the decisions that directors make to create physicalized performance of text for deeper understanding. We need to encourage discussions of climate change not just as a "fun" activity but as serious academic work incorporating artistic imagination as well as scientific investigation. While teachers may be tempted to ignore this passage as just a long monologue about bad weather, once students spend some time with it, they may see it as much more penetrating in its imagination of climate disaster. In writing about Titania's speech, Homer Swander asked,

> How likely is survival on a planet that is under attack from revenging winds and an angry moon? How likely is survival when the only air to breathe is ripe with disease? How likely is survival when it is impossible to know what season it is today or will be tomorrow? How likely is survival on a planet where nature itself has become unnatural, permanently so? For if the fairies never dance, the evils will last forever.[64]

These questions that Titania requires us to ask in the context of Shakespeare's 400-year-old text are commensurate with the very questions we must ask ourselves now as we attempt to understand and act upon our understanding of climate change.

The workshop is not without its limitations. One problem is the privileging of one reading of Shakespeare's text over possible others. Indeed, the premise of the workshop itself hinges on our accepting Titania's anthropomorphization as an indication that she somehow represents humanity, an interpretation that not all students may share. Doing so also limits discursive space for other interpretations of her character, either as representing nature itself or simply as a representative of a faerie world that was once commonly believed to exist in reality before Christianity drove the fay folk out of England. Furthermore, connecting the monologue to various natural disasters presupposes the veracity of Titania's claim within the world of the play, a presupposition not always maintained by directors or scholars. The disastrous weather patterns mentioned in the speech do not appear to impact the action of the play going forward. The fog that obscures Lysander from Demetrius in their fight scene in Act 3 of the play is conjured by Puck, through Oberon's wish to remove the "error" (3.2.369) that led to the fight, not through the dissension between the royal faerie couple. While connections may be made between the weather described in the speech and the particularly cold, wet weather in England in the 1590s, *MND* is set in Athens and thereabout, hardly a cold, wet climate. Another problem rests in the process of developing a performance, as one performance choice suggests others and progressively limits the field of options available in constructing a scene that has any sense of unity. As English teachers, we want to suspend interpretive closure for as long as possible; therefore, the performance suggestions included in this chapter are examples, not maxims, though by virtue of their inclusion, they may limit other and competing, or at least different, interpretations. If teachers can contextualize the workshop by acknowledging the limitations and presuppositions inherent in the workshop, students may feel empowered to offer alternative and unanticipated readings of the speech.

In essence, creating space in class for theater practices to engage students with literature and with current events engenders artistic lens-building for students who dearly need aesthetic consciousness and broad imagination to negotiate what is likely to become an increasingly scary and disjointed world. One worry many of us share is that, as climate change continues to disrupt our lives, and does so with greater intensity, youth may lose hope in the dream of a better future, or indeed any future at all. We must address climate change across the curriculum, but we must do so with care and an undying sense of hope. Theater, no matter how serious the topic, not only broadens our perspectives, but also increases our sense of creativity, energy, and fun. Children are emerging from the most disruptive chapter of their collective lives, thanks to the COVID-19 pandemic, which may be only a precursor to decades of climate-related catastrophes that they will have to manage in adulthood. If one goal of education is to equip an enlightened (global) citizenry to build toward an equitable, just, and sustainable future, then we need to be brave in our choices of classroom topics and creative in our choices of methods. Our children need and deserve our best in this regard.

NOTES

1. William Shakespeare, *A Midsommer Nights Dreame*. First Folio edition (London, 1623).
2. My thanks go to Marilyn Halperin, Jason Harrington, and Molly Truglia of the education team at Chicago Shakespeare Theater, where this workshop was first conducted.
3. Line markers are from Stephen Greenblatt, *The Norton Shakespeare, Volume 1: Early Plays and Poems*, 2nd ed. (New York: Norton, 2008), 858–859.
4. Richard Beach, Jeff Share, and Allen Webb, *Teaching Climate Change to Adolescents: Reading, Writing, and Making a Difference* (New York: Routledge, 2017).
5. Beach, Share, and Webb, *Teaching Climate Change to Adolescents: Reading, Writing, and Making a Difference*, 9.
6. U.S. Global Change Research Program, *Climate Science Special Report: Fourth National Climate Assessment, Volume I* (Washington, DC: U.S. Global Change Research Program, 2017).
7. Laura Barbas-Rhoden and Bethany Wiggins, "A Message from the ASLE Leadership," accessed January 27, 2021, https://www.asle.org/discover-asle/vision-history/.
8. Barbas-Rhoden and Wiggins.
9. Gabriel Egan, *Green Shakespeare* (Abington, Oxfordshire: Routledge, 2006).
10. Egan, *Green Shakespeare*, 124.
11. Egan, *Green Shakespeare*, 124.
12. Egan, *Green Shakespeare*, 124.
13. Shakespeare Association of America. "January Bulletin," Last modified January 1, 2021, https://shakespeareassociation.org/wp-content/uploads/2021/01/January-Bulletin-2021.pdf.
14. Simon C. Estok, "Coda: Ecocriticism on the Lip of a Lion," in *Ecocriticism and Shakespeare: Literatures, Cultures, and the Environment* (New York: Palgrave Macmillan, 2011), 124.
15. Rebecca Bushnell, "Shakespeare and Nature," in *Shakespeare in Our Time: A Shakespeare Association of America Collection*, eds. Dympna Callaghan and Suzanne Gossett (London: Bloomsbury, 2016), 327–328.
16. E. M. W. Tillyard, *The Elizabethan World Picture* (New York: Vintage, 1959).
17. Tillyard, *The Elizabethan World Picture*, 61.
18. Tillyard, *The Elizabethan World Picture*, 26.
19. Steve Mentz, "Shakespeare Without Nature," in *Shakespeare in Our Time: A Shakespeare Association of America Collection*, eds. Dympna Callaghan and Suzanne Gossett (London: Bloomsbury, 2016), 338.

20. Naglaa Abu-Agog, "Ecocriticism and Shakespeare: The Gaia Hypothesis in As You Like It," *International Journal of Studies in English Language and Literature* 4, no. 5 (2016): 14.

21. Greenblatt, *Shakespeare*, 2.1.81–117.

22. *A Midsummer Night's Dream*, directed by Max Reinhardt and William Dieterle (Burbank, CA: Warner Bro. Pictures, 1935).

23. *A Midsummer Night's Dream*, directed by Adrian Noble (Stratford-upon-Avon: Royal Shakespeare Company; Burbank, CA: Buena Vista Home Entertainment, 1996).

24. *A Midsummer Night's Dream*, directed by Michael Hoffman (Burbank, CA: Fox Searchlight Pictures, 1999).

25. Teaching Shakespeare Institute, *Shakespeare Set Free: Teaching A Midsummer Night's Dream, Romeo and Juliet, and Macbeth*, ed. Peggy O'Brien (New York: Washington Square Press, 1993).

26. Teaching Shakespeare Institute, 58–60.

27. Ralph Alan Cohen, *ShakesFear and How to Cure It: A Handbook for Teaching Shakespeare* (Clayton, ED: Prestwick House, 2006), 231.

28. *A Midsummer Night's Dream*, directed by Peter Hall (Sonoma, CA: Filmways and Stratford-upon-Avon; Warwickshire, UK: Royal Shakespeare Enterprises, 1968).

29. *A Midsummer Night's Dream*, directed by Julie Taymor (London: Ealing Studios/Londinium, 2014).

30. *A Midsummer Night's Dream*, directed by Casey Wilder Mott (Los Angeles, CA: Brainstorm Media/Emperian Productions/5G Studios, 2018).

31. Tripthi Pillai, "Constructing Experiences and Charting Narratives," in *A Midsummer Night's Dream: A Critical Guide*, ed. Regina Buccola (London: Continuum, 2010), 153.

32. Pillai, "Constructing Experiences and Charting Narratives," 154.

33. Sophie Chiari, "Climate as Climax in Shakespeare's Plays," *Shakespeare in Southern Africa* 29 (2017): 3.

34. Chiari, "Climate as Climax in Shakespeare's Plays," 3–4.

35. Milla Cozart Riggio, "Introduction" in *Teaching Shakespeare through Performance*, ed. Milla Cozart Riggio (New York: The Modern Language Association of America, 1999), 5.

36. Homer Swander, "In Our Time: Such Audiences We Wish Him," *Shakespeare Quarterly* 35 (1984): 529, quoted in David Kennedy Sauer and Evelyn Tribble, "Shakespeare in Performance: Theory in Practice and Practice in Theory," in *Teaching Shakespeare through Performance*, ed. Milla Cozart Riggio (New York: The Modern Language Association of America), 35–36.

37. Edward Rocklin, *Performance Approaches to Teaching Shakespeare* (Urbana, IL: National Council of Teachers of English), 22.

38. Milla Cozart Riggio, *Teaching Shakespeare through Performance*, ed. Milla Cozart Riggio (New York: The Modern Language Association of America, 1999).

39. Teaching Shakespeare Institute, *Shakespeare Set Free: Teaching A Midsummer Night's Dream, Romeo and Juliet, and Macbeth*, ed. Peggy O'Brien (New York: Washington Square Press, 1993).

40. Timothy J. Duggan, "Drama Workshop in the English Classroom: Studying Shakespeare through the Eyes of Actors and Directors," *Shakespeare* 10, no. 4 (2003): 10.

41. Richard Beach, Jeff Share, and Allen Webb, *Teaching Climate Change to Adolescents: Reading, Writing, and Making a Difference* (New York: Routledge, 2017), 100.

42. If the class is working from small, paperback scripts, they may be able to use those without requiring a photocopied text of the speech.

43. Class size may require that teachers cut parts of the speech to assure that students are able to collaborate with their peers to create these enactments. For example, cutting four of the beats would allow a class of twenty-four students to form eight groups of three students.

44. Greenblatt, *Shakespeare*, 2.1.88–90.

45. Greenblatt, *Shakespeare*, 2.1.90–92.

46. Greenblatt, *Shakespeare*, 2.1.93–95.

47. Greenblatt, *Shakespeare*, 2.1.96–97.

48. Greenblatt, *Shakespeare*, 2.1.98–100.

49. Folio and Quarto versions of the script say "here," but Norton 2nd ed. uses "cheer" as the more likely word. See Stephen Greenblatt, *The Norton Shakespeare, Volume 1: Early Plays and Poems*, 2nd ed. (New York: Norton, 2008), 859.

50. Greenblatt, *Shakespeare*, 2.1.101–102.

51. Greenblatt, *Shakespeare*, 2.1.103–105.

52. Greenblatt, *Shakespeare*, 2.1.106–108.

53. Greenblatt, *Shakespeare*, 2.1.109–111.

54. Greenblatt, *Shakespeare*, 2.1.111–113.

55. Greenblatt, *Shakespeare*, 2.1.113–114.

56. Greenblatt, *Shakespeare*, 2.1.115–116.

57. Greenblatt, *Shakespeare*, 2.1.117.

58. National Governors Association Center for Best Practices and Council of Chief State School Officers, *Common Core State Standards for English Language Arts and Literacy in History/Social Studies, Science, and Technical Subjects* (Washington, DC: National Governors Association Center for Best Practices and Council of Chief State School Officers, 2010), 4–5.

59. Greenblatt, *Shakespeare*, 2.1.88–90.

60. American Lung Association, "Toxic Air Pollutants," updated July 13, 2020, https://www.lung.org/our-initiatives/healthy-air/outdoor/air-pollution/toxic-air-pollutants.html.

61. Thanks to Marilyn Halperin for helping to develop this idea.

62. Greenblatt, *Shakespeare*, 2.1.88–90.

63. The lesson plan that accompanies this chapter includes links to short informational texts corresponding to each segment of Titania's monologue.

64. Homer Swander, *Dreame: A New Play by Shakespeare*. Self-published manuscript (2012): 97–98.

REFERENCES

Abu-Agog, Naglaa. "Ecocriticism and Shakespeare: The Gaia Hypothesis in As You Like It." *International Journal of Studies in English Language and Literature* 4, no. 5 (2016): 1–8.

American Lung Association. "Toxic Air Pollutants." *American Lung Association* Website, American Lung Association, updated July 13, 2020, https://www.lung.org/our-initiatives/healthy-air/outdoor/air-pollution/toxic-air-pollutants.html

Barbas-Rhoden, Laura and Bethany Wiggin. "A Message From the ASLE Leadership," *ASLE* Website, Association for the Study of Literature and Environment, Accessed on January 31, 2021, https://www.asle.org/discover-asle/vision-history/

Beach, Richard, Jeff Share, and Allen Webb. *Teaching Climate Change to Adolescents: Reading, Writing, and Making a Difference.* New York: Routledge, 2017.

Boal, Augusto. *Theater of the Oppressed.* New York: Theater Communications Group, 1993.

Bushnell, Rebecca. "Shakespeare and Nature." In *Shakespeare In Our Time: A Shakespeare Association of America Collection*, edited by Dympna Callaghan and Suzanne Gossett, 327–334. London: Bloomsbury, 2016.

Chiari, Sophie. "Climate as Climax in Shakespeare's Plays." *Shakespeare in Southern Africa* 29 (2017): 1–15.

Cohen, Ralph Alan. *ShakesFear and How to Cure It: A Handbook for Teaching Shakespeare.* Clayton, DE: Prestwick House, 2006.

Duggan, Timothy J. "Drama Workshop in the English Classroom: Studying Shakespeare through the Eyes of Actors and Directors." *Shakespeare* 10, no. 4 (2003): 10–12.

Egan, Gabriel. *Green Shakespeare: From Ecopolitics to Ecocriticism.* Abington, Oxfordshire: Routledge, 2006.

Estok, Simon C. "Coda: Ecocriticism on the Lip of a Lion." In *Ecocriticism and Shakespeare: Literatures, Cultures, and the Environment*, 123–125. New York: Palgrave Macmillan, 2011.

Greenblatt, Stephen. *The Norton Shakespeare, Volume 1: Early Plays and Poems* (2nd ed.). New York: Norton, 2008.

Hall, Peter, dir. *A Midsummer Night's Dream.* Sonoma, CA: Filmways and Stratford-Upon-Avon; Warwickshire, UK: Royal Shakespeare Enterprises, 1968.

Hoffman, Michael, dir. *A Midsummer Night's Dream.* Burbank, CA: Fox Searchlight Pictures, 1999.

Kennedy Sauer, David and Evelyn Tribble. "Shakespeare in Performance: Theory in Practice and Practice in Theory." In *Teaching Shakespeare through Performance*, edited by Milla Cozart Riggio, 33–47. New York: The Modern Language Association of America, 1999.

Mentz, Steve. "Shakespeare Without Nature." In *Shakespeare In Our Time: A Shakespeare Association of America Collection*, edited by Dympna Callaghan and Suzanne Gossett, 334–338. London: Bloomsbury, 2016.

Mott, Casey Wilder, dir. *A Midsummer Night's Dream*. Los Angeles, CA: Brainstorm Media/Emperian Productions/5G Productions, 2018.

National Governors Association Center for Best Practices and Council of Chief State School Officers. *Common Core State Standards for English Language Arts and Literacy in History/Social Studies, Science, and Technical Subjects*. Washington, DC: National Governors Association Center for Best Practices and Council of Chief State School Officers, 2010.

Noble, Adrian, dir. *A Midsummer Night's Dream*. Stratford-upon-Avon: Royal Shakespeare Company and Burbank, CA: Buena Vista Home Entertainment, 1996.

Pillai, Tripthi. "Constructing Experiences and Charting Narratives." In *A Midsummer Night's Dream: A Critical Guide*, edited by Regina Buccola, 150–171. London: Continuum, 2010.

Raber, Karen. "The Chicken and the Egg." In *Shakespeare In Our Time: A Shakespeare Association of America Collection*, edited by Dympna Callaghan and Suzanne Gossett, 338–341. London: Bloomsbury, 2016.

Reinhardt, Max and William Dieterle, dirs. *A Midsummer Night's Dream*. Burbank, CA: Warner Bros. Pictures, 1935.

Riggio, Milla Cozart. *Teaching Shakespeare through Performance*. New York: The Modern Language Association of America, 1999.

Riggio, Milla Cozart. "Introduction" In *Teaching Shakespeare through Performance*, edited by Milla Cozart Riggio, 1–18. New York: The Modern Language Association of America, 1999.

Rocklin, Edward L. *Performance Approaches to Teaching Shakespeare*. Urbana, IL: National Council of Teachers of English, 2005.

Shakespeare Association of America. "January Bulletin," Shakespeare Association of America Website, Shakespeare Association of America, Last modified January 1, 2021. https://shakespeareassociation.org/wp-content/uploads/2021/01/January-Bulletin-2021.pdf

Swander, Homer. *Dreame: A New Play by Shakespeare*. Self-published manuscript (2012) In the author's possession.

Taymor, Julie, dir. *A Midsummer Night's Dream*. London: Ealing Studios/Londinium, 2014.

Teaching Shakespeare Institute. *Shakespeare Set Free: Teaching A Midsummer Night's Dream, Romeo and Juliet, and Macbeth*, edited by Peggy O'Brien. New York: Washington Square Press, 1993.

Tillyard, E.M.W. *The Elizabethan World Picture*. New York: Vintage, 1959.

U.S. Global Change Research Program. *Climate Science Special Report: Fourth National Climate Assessment, Volume I*. Washington, DC: U.S. Global Change Research Program, 2017.

Chapter 7

Raising Environmental Awareness and Language Acquisition through Haiku

Lorraine Kerslake and María Encarnación Carrillo-García

The use of haiku as a creative form of writing and a literary tool for environmental education can be implemented in the classroom to raise awareness and appreciation of nature by encouraging a more sustainable and respectful relationship with the natural world. In a world language classroom teachers can implement a didactic methodology where haiku can become an educational resource for learning while promoting critical environmental values.

Such concerns may seem less pressing at times like the present when the world as we knew it belongs to the past and we are facing a global pandemic that is transforming the society in which we live. However, the present crisis is part of a much broader problem, one deeply connected to our dysfunctional relationship with nature. Our disruption of ecosystems and human exploitation of other species, including encroachment on wild animals' habitats, has caused the emergence of spillover diseases such as SARS and COVID-19, the most devastating virus to date in modern history.[1]

But what if we viewed this pandemic as a teaching moment? Now, more than ever, imagination is key in envisioning change and rethinking the way we interact with the world around us. Our culture is predominately anthropocentric, one which considers humankind as superior to nature and other animals and one that is guided by the assumption that it is only in relation to humans that anything else has value. As Val Plumwood claims, this "ecological mess" that we have made is the

> fruit of a human and reason-centred culture that is at least a couple of millennia old, whose contrived blindness to ecological relationships is the fundamental condition underlying our destructive and insensitive technology and behaviour. To counter these factors, we need a deep and comprehensive restructuring of

> culture that rethinks and reworks human locations and relations to nature all the way down.[2]

According to Plumwood, restructuring our culture requires a deep critical awareness of the discursive structures that construct the reality we live in. Traditionally, language, as a human construct, is aligned with culture and is therefore positioned against nature, so how can we "rewrite" the world and engage with the relations between human and nonhuman nature? And, more specifically, as educators, how can we raise environmental awareness in the classroom and get students to engage and empathize with the world around them?

TOWARD A HUMANISTIC APPROACH TO LANGUAGE TEACHING: GLOBAL EDUCATION AND THE ROLE OF LITERATURE

There is need in our current world for a more humanistic education, as Peter Grundy[3] reminds us. That is to say, a humanistic, or holistic, way of teaching, or education, of the whole person, with a strong emphasis on the self-motivation of the learner in order to foster the individual's sense of self-esteem while respecting the student's own knowledge and independence. This humanistic educational approach takes into consideration the affective character of language learning but also requires a rethinking of traditional syllabuses and materials in the classroom. By creating a more personal learning context in which to use these educational strategies both inside and outside the classroom we encourage students to use language freely and more independently.

As educators we believe that all teaching methods should be personalized according to the group of learners, making the personal experience of learning the first educational resource for all lesson contents. Promoting new ways of critical thinking by equipping educators and learners with relevant competencies has been the purpose of many pedagogic theories that precede the concept of global education, which was conceived as a holistic education adapted to the twenty-first-century world context of globalization. As such, globalization has affected educational reforms and global citizenship education is now being implemented in classrooms in many countries in Latin American, North America, Europe, Africa, the Middle East, and Asia. Because of this, global education has emerged as a pedagogical approach and an attempt to combine diverse methodological approaches and bring them into the curriculum by promoting learners' world knowledge about people, countries, cultures, and global issues such as peace, development, the environment, or human rights.[4] According to the Maastricht Global Education Declaration, "Global education is education that opens people's eyes and minds to the realities of the globalised world and awakens them to bring about a world of greater justice, equity and Human Rights for all."[5] As an approach to language teaching, global education aims to integrate this holistic perspective into the classroom by promoting critical thinking.

As stated by Kip Cates, a global education approach in language studies involves "integrating a global perspective into classroom instruction through a focus on international themes, lessons built around global issues, classroom activities linking students to the wider world and concepts such as social responsibility and world citizenship."[6] The integration of environmental issues in the languages classroom can contribute to developing a more humanistic and global way of understanding and interacting with the world. As educators we should be concerned not only with thinking about but also with creating language in a holistic way while recognizing the connection between humans and nature. In order to implement this educational perspective in the classroom, it is advisable for teachers to start from a defined methodological approach that gives them the necessary support to develop the class sessions. Following this, our didactic proposal has been developed considering the "Topic-tasked based syllabus" approach by James Bourke[7] in which the topic selected by the teacher to develop the activities is the environment, and in which the activities proposed are related to the reading and writing of poems.

In relation to the teaching of languages other than the mother tongue, we must highlight that the theoretical tradition in language teaching distinguishes between the teaching of foreign languages and the teaching of second languages. The Oxford Advanced American Dictionary[8] defines the word foreign as a reality from a country that is not your own; in this sense, authors such as Muñoz[9] explain that a difference is made between these two terms to emphasize that, in the case of so-called second languages, it is a language spoken in the community where the learner lives, even if it is not his or her mother tongue, while in the case of foreign languages, the language does not have a presence in the community where the learner lives. For example, English is a second language for a Mexican immigrant in the United States, while it is a foreign language for a Spanish student in Spain. Therefore, in this chapter we address foreign language teaching because the learning context we are describing is English language instruction for higher education students in Spain.

From primary to tertiary education, literature is listed in the curriculum as one of the tools that can be used for the successful implementation of sociolinguistic, sociocultural, and emotional aspects of the teaching of English as a Foreign Language (EFL). Following this, it is important to bear in mind the role that literature, as an authentic resource, can have in fomenting a humanistic conception of the syllabus.[10]

Literature is a powerful tool for studying languages and can help to foster global education by considering the emotions of the learner, since it uses a kind of connotative language that stimulates our soul and intellect. Also, it refers to important matters about human existence while turning classroom activities into an emotional experience that can help with language learning.

In this context, poetry can become a powerful resource for studying other languages because of its connotative nuances in diction that stimulate the soul and intellect. Not only does poetry communicate important matters about the human existence but it can also turn classroom activities into an emotional experience that can foster learning. In this sense, Francisco Mora,[11] a neuroscientist, claims that emotion and cognition, both mental processes, form an indissoluble binomial where the concepts learned are impregnated by our emotions, that is to say that the brain only learns if there is emotion. This "passion for learning" is particularly relevant for teachers who wish to give emotional experiences to their students while learning.

As Whitman and Kelleher describe, neuroscience has demonstrated that when arts are integrated into the everyday teaching methodology in other nonartistic areas of the curriculum, they can help create enduring learning and also increase student motivation. As they explain, "Students need opportunities to transfer their knowledge through the visual and performing arts, so integrate art into the core content of no-arts subjects to enhance learning."[12] In this sense, Rinne, Gregory, Yarmolinskaya, and Hardiman also highlight that arts integration improves long-term retention of content, suggesting that it creates knowledge transfer by forcing students to synthesize and rearrange ideas, and is linked to long-term memory consolidation and engagement.[13]

David Sousa's work[14] also supports the idea that the arts play an important role in human development, fostering the development of cognitive, emotional, and psychomotor pathways.

Since from this perspective literature can be understood as a work of art, and thus be a source for learning another language, then as Peter Chofer[15] asserts, the most accessible texts can be models not just for reading but also for performing language. In this way, it would be a good educative practice to establish literature as an educational resource from which to work on the contents of language and literature, in itself, as well as to provide the opportunity to work on the contents from an emotional and sensory point of view. This favors the integral learning of concepts from both cerebral hemispheres[16] in a holistic and humanistic way. Following this, literature, and in particular, poetry, in our case haiku, can be a useful tool because it allows us, as teachers, the opportunity to learn the language through a work of art, the haiku, and to offer real emotional experiences to our students inside the classroom.

WHY HAIKU?

Working with literature as a form of meaningful literacy in the EFL classroom, as David Hanauer[17] defends, can give students the opportunity to learn independently about the presence of others, and from experiences that connect them to their inner self and to the natural world, thereby promoting a more humanistic or all-inclusive way of teaching. When thinking about the practices of literacy for the purpose of our study we are particularly interested in drawing on the concept of environmental literacy or eco-literacy, as outlined by David W. Orr. As Orr correctly asserts, "All education is environmental education. By what is included or excluded, emphasized or ignored, students learn that they are part of or apart from the natural world."[18] In our analysis we propose the use of haiku as a valid element for education and a literary tool for environmental awareness. Haiku can also be understood as a form of universal poetry, implemented in the classroom as an approach to creative writing and as an effective way to encourage learners to practice English as a language in use, that is, as one which appears in real-life contexts other than in textbooks. In doing so we are also helping our pupils to use EFL by living it and by drawing on emotional, linguistic, literary, cultural, and critical aspects, within a specific context—in other words, learning by using.

A haiku is a traditional form of Japanese poetry inspired by nature—landscapes, wildlife, and waterscapes that belong to the natural world we live in. Based on a system of three lines it has a set syllable pattern where the first and last lines have five syllables, and the middle line has seven syllables. The simplicity of the seventeen syllables condensed in its 5-7-5 pattern works as a structure around which a thought is often written in one inspired moment that lingers in the imagination. The point of a haiku is implied rather than directly expressed and so haiku have been styled "the half-said thing."[19] Rather than being told what to think and feel, haiku stress particular importance on the here and now, allowing the reader to share a moment of nature and participate in the experience.

This can be seen in one of the greatest haiku writers, the Japanese poet Basho (1644–1694), who expresses this sense more directly and revolutionized haiku with his "frog" poem. What makes the verse striking in its time is the fact that Basho was the first poet to present a frog not croaking but, instead, to focus on the sound of the frog jumping into the water:

An old pond; / a frog jumps in—/ the sound of the water.

Basho's sense of appreciation can be inferred through the nature of the discourse of haiku. Most animals represented in haiku are not doing anything extraordinary but act as simple agents doing what they would normally do.

This is a real frog before the eyes of the poet, not an imaginary one. The appreciation of the ordinary and the emphasis on the present moment allows the poet to capture and imagine it from the other's point of view. This act of empathy, the ability to identify with the other, is also key to creating a caring relationship with nature, and with the development of a new and necessary environmental awareness that must be part of the concept of global citizenship, especially in the face of climate change.

Basho's real name was Matsuo Kinsaku, but he changed his name to Basho in honor of a banana tree, rare to Japan, given to him by one of his disciples. Basho was a student of Zen Buddhism who converted even the most trivial things into poetry. He wrote over 1,000 haiku and lived a simple life in which he "gave up virtually all possessions, his only concern spiritual and artistic discovery."[20]

After Basho, Yosa Buson (1715–1783) is the second great master of Haiku. Buson gained his income not from haiku but primarily from landscape painting, which can be experienced in the strong sensory images depicted in the following poem:

Willow leaves fallen / Clear waters dried up stones / One place and another

The autumn landscape depicts the leafless willow hanging over the stones of the stream creating for the reader a romanticized visual conceived by the artist in the reader's mind. By becoming an agent of the poem, the poet also becomes a painter. In effect, a haiku is understood primarily as *shasei*, that is, as a sketch of nature.[21] In the case of our students, the image that is conceived when reading and writing haiku can also help to develop environmental awareness while learning a language, bringing them closer to the natural world due to the great connection between the traditional theme of haikus, nature, in general, and seasons, in particular.

Ever since Massaoka Shiki (1867–1902) advocated that "haiku is literature,"[22] it has been considered one of the most important traditional literary forms in Japan. Shiki modernized the traditional form of haiku as a means for artistic expression and took up the practice of going out into nature to craft his work, which he considered poetic sketches characterized by evocative visual descriptions.

Yoel Hoffman characterized the Japanese relationship with nature as "no mere gesture of aesthetic appreciation, but an act of worship,"[23] referring to the capacity of the poet to bring us back to Earth and reconnect with the natural world. This idea of the poet creating a sensory affinity between the reader and nature to nurture a sense of wonder is very much in line with what Rachel Carson described in her essay "Help Your Child to Wonder,"[24] in which she reminds us that we are all part of the natural world and urges us to explore nature with all our senses.

David Abram[25] argues that the invention of writing is largely responsible for disconnecting humans from the rest of the living world. Abram is concerned with the impact of written language. Not only does he believe that language can alienate us from the natural world, but he also draws on the capacity of the writer to use language to reconnect us and bring us back to Earth. For this purpose, he proposes a new alternative discourse that reflects and evokes nature. As he eloquently puts it,

> There can be no question of simply abandoning literacy, of turning away from all writing. Our task, rather, is that of *taking up* the written word, with all of its potency, and patiently, carefully, writing language back into the land. Our craft is that of releasing the budded, earthy intelligence of our words, freeing them to respond to the speech of the things themselves—to the green uttering-forth of leaves from the spring branches.[26] Emphases in original.

Haiku can dissolve the boundaries between the disconnected human mind and nature and help in this way to bring us back to Earth. As noted by Arran Stibbe,

> The importance of the discourse of haiku is that, like a Zen *koan*, it uses language to encourage the reader to go beyond language, beyond the world of intellectual abstractions, and reconnect directly with the more-than-human world. The way it does this is to describe actual encounters with everyday nature, in straight present tense, using a minimal amount of metaphor and abstraction, placing poetic emphasis on individual animals and plants, representing them as agents of their own lives living according to their natures, with implicit assumptions of empathy and positive regard built into the discourse.[27]

By creating empathy and showing respect toward animals and nature, haiku go beyond language and express an appreciation, or what Carson called a sense of wonder, toward the natural world they describe. By modeling a close respectful relationship of humans with nature they allow the reader to value and appreciate the other by encouraging a form of direct ecological consciousness.

Nowadays haiku can be considered as a kind of world poetry. They no longer follow established patterns and are composed in many languages beyond Japanese. However, it is important to recognize the value and know the techniques of traditional haiku to capture a moment through descriptive and visual verse.

The brevity of a haiku demands a very active participation from the reader in reading and forming a mental image while pondering over the meaning of the descriptive verse before their eyes. This means that as a teacher it is no good telling students that a haiku is simply a poem about nature written in seventeen syllables. Haiku focus on the here and now, forcing us to slow down and "be in the moment," like an active form of meditation. This does not mean we have to teach our students to become Zen Buddhists or sit with crossed legs but instead to have a certain attitude of awareness. A more practical hands-on approach would be to take students outdoors and go for a walk, asking them to use their senses to observe nature and then to write a haiku.

This type of creative practice integrates environmental education and learning in the classroom to foster a more holistic approach. Environmental education goes beyond simply reading about the environment. Orr's idea of ecological literacy understood as "the capacity to observe nature with insight"[28] is particularly useful here and implies looking at how environmental education goes beyond simply reading about nature to engaging with it in an insightful and interactive way. Ecological literacy places humans as integral parts of ecosystems and recognizes the impact of our actions on other species of the planet. Orr advocates a form of education that will change not only our anthropocentric vision of the reality but one that cultivates our ecological imagination. In *Earth in Mind*, he further develops the concept of ecological literacy by stating that

> the disordering of ecological systems and of the great bio-geochemical cycles of the earth reflects a prior disorder in the thought, perception, imagination, intellectual priorities and loyalties inherent in the industrial mind. Ultimately, the ecological crisis concerns how we think and the institutions that purport to shape and refine the capacity to think.[29]

In other words, Orr believes that the intellectual failure to understand the environmental crisis stems largely from the failure in our educational system when environmental values are not worked on in the classroom in an integrated way throughout the curriculum. Following this, education must take a key role in redefining our vision of reality in order to cultivate environmental intelligence as well as ecological imagination.

Therefore, it is our task as educators and teachers not only to develop critical and creative thinking in the classroom but, by extending Orr's ecological literacy, to raise awareness and contribute toward an ecological consciousness. By recognizing the importance of these issues and incorporating them into the classroom we can raise awareness for increased engagement and empathy with the natural world. However, reading and writing haiku not only requires an appreciation toward nature but also requires the ability to listen to nature. Haiku uses language to encourage the reader to go beyond language. In Orr's words,

> The language of nature includes the sounds of animals, whales, birds, insects, wind, and water: a language more ancient and basic than human speech. To hear this language requires patient, disciplined study of the natural world. But it is a language for which we have an affinity.[30]

This affinity toward nature requires recognizing animals (and plants) as subjects and agents. If haiku involve listening to nature, they also entail empathy on behalf of the reader in order for them to imaginatively project themselves into a different perspective and to imagine the world from the other's point of view. Furthermore, as will be revealed, when ecological and environmental concerns are brought together haiku are particularly effective in helping to challenge our perceptions and to shift them from anthropocentric to biocentric.

Haiku in Language Instruction

The idea of using haiku in the language classroom is not new. Bearing in mind that our learning experience takes place in a classroom where students are learning another language, we consider it of interest to briefly review other authors who have highlighted the value of haiku in this circumstance.

In the context of Japan, Atsushi Iida's research includes looking at how language learners express their own voice through creative writing and haiku writing pedagogy. Iida[31] highlights that Haiku poetry in an English composition class can provide students with opportunities to rediscover their "self" through the process of writing while developing linguistic and cultural sensitivities of English. One of the activities he details[32] is that where students each create a book of haiku as one of the course requirements. Iida also carries out research related to the assessment tools for haiku poetry in the EFL classroom and focuses his attention on the evaluation of the form (lines, pattern 5-7-5 syllables, descriptive language) and the content (integration of nature and human mind, personal voice) and creates an analytic rating scale. This scale serves to check features such as personal voice, audience awareness, organization, haiku, and linguistic conventions where haiku creations can be marked as excellent, fair, or poor.

Maria Moss presents an insightful article based on her own experience teaching creative writing in language classes at Leuphana University in Lüneburg. "Writing Creatively in a Foreign Language: Vignettes, Haikus, and Poetry" focuses on writing materials and strategies used in her "step-by-step approach" to develop the students' ability of writing creatively. The author describes the creative tasks the students perform, such as haiku, and concludes that "writing Haikus is a wonderful way of creating an awareness of syllables, sound units, and the management of both—its rhythms and pauses, its vowel and consonant structure."[33]

Yong Lee[34] also looks at EFL education in universities in Japan and the effectiveness of poetry and in particular the use of haiku writing in the context of EFL. Drawing on Swain,[35] he shows how haiku writing can be effective in the language acquisition process. To do so he describes the three functions of output, noticing, hypothesis-testing, conscious reflection, and applies them to the experience of haiku writing in the EFL classroom. First, students will discover whether their linguistic knowledge is sufficient to express their thoughts and feelings; second, they practice with the language in an experimental way helping with the consolidation of the language; finally, they reflect on language through the work carried out by choosing words and phrases when creating haiku. As he concludes, haiku writing is particularly effective in this context given that the language production by the learners becomes easier to remember for them, thus contributing to language acquisition. We can consider these writers and their pedagogical experiences as forerunners and inspirers of the research project and educational experience presented by the authors.

THE POETRY READING PROJECT: WRITING HAIKU, NATURE, AND ENVIRONMENTAL EDUCATION

If we cannot be creative teachers able to produce activities that engage our students, we cannot expect them to be motivated, given the scarcity of literary models for creative writing in EFL classrooms, and the lack of activities that focus on creativity inside the official curriculum. Bearing this in mind, we decided to develop, in collaboration, two educational experiences based in two different Spanish universities: The University of Alicante and the University of Murcia, both situated in the southeast on the Mediterranean coast with students who studied EFL. The experience at the University of Alicante was a pilot for the project[36] that would later be developed at the University of Murcia during the first and most difficult months of the pandemic in 2020.

We wish to highlight that the seeds of the haiku experience began in a language class on English culture and literature given in a master's course of English and Spanish as Foreign Languages at Alicante University after asking students about the importance of teaching poetry in the EFL classroom, including its possible benefits and drawbacks.

Among the benefits listed students claimed that poetry could help with language acquisition (pronunciation, phonetic awareness, rhyme, intonation) and vocabulary, given that most poems are short. A few more enthusiastic students even suggested that poetry could be fun (although others looked rather skeptical at this claim). What concerned us the most though were their responses for its drawbacks. The majority said they were more worried about finishing their syllabus and teaching program and that they simply did not have time to "play with poetry." Others could not see the point; after all, if their students were not interested in reading poetry in their mother tongue why would they want to read it in English? Even those who had tried using poetry claimed that it was hard enough for them to understand a poem without trying to explain it to their students, most of whom only wanted to be told what the poem was about. This led to the awareness that it was necessary to initiate these graduate students in a simple and creative way to teaching poetry: through haiku. This was put into practice in the next class by showing a series of haiku and giving students the opportunity to discover for themselves that even they could become poets for a day. Among teachers who use haiku one of the controversial issues is whether or not to insist on the use of traditional conventions, including a strict syllable count of seventeen syllables 5-7-5. Although this should not be deemed compulsory, given that in this case the majority of students were learners with a high language proficiency,[37] they were encouraged to respect the tradition together with three other "golden rules" as guidelines about what a haiku should do:

- It captures the essence of a moment in time: a sort of poetic photo-shot.
- It finds the extraordinary in the ordinary: imagine you are looking at nature for the first time.
- It reconnects our world to nature.

The next task was to choose a topic related to a particular element of nature. It is important to encourage students to choose an ecosystem of interest to them when constructing their haiku. Given that we live next to the Mediterranean Sea it seemed logical to choose this as inspiration. We then brainstormed a shortlist of words (ten to fifteen) related to the theme. Students were asked to write down the first thing that came to their mind. In this case their lists included words such as sea, salt, beach, wave, jellyfish, swimming, sand, happiness, splash, water, blue.

As the class was made up of about twenty students they were then put in pairs and asked to write down the number of syllables next to each word. Given the number of students this activity was easy to monitor while in progress. They were then asked to draft their haiku following the traditional pattern and share these with the rest of the class. The following are just a few examples of the results:

Above the blue sea / There's a jellyfish waiting / For a kid to sting.

The sea: calm and blue / Water takes your stress away / Then wrinkled fingers.

The sea and the sun / Sandy hands sink into waves / Splashy, salty scene.

This simple exercise illustrates not only how easy the task is but also how poetic the results can be and is only one suggestion for written practice using haiku. Being only three lines long, haiku is easy for all students to write in a second language, regardless of their level and makes it possible for learners to express their thoughts and emotions freely and creatively.

In view of the results and positive feedback from the students who participated and following an exchange of ideas related to the poetry reading project, we decided to carry out an experimental workshop related to using poetry and haiku in the language classroom at the University of Murcia. We proposed a teaching project entitled "The Poetry Reading Project." The project, which began before the pandemic and ended inside it, was selected by the Innovation Unit at the University of Murcia (Spain)[38] and developed during the academic year 2019 to 2020, among a group of fifty-seven students who were studying a Degree in Education, in particular for the subject EFL for Infant Education.

To develop the project we followed a linguistic, literary, and creative approach as described by Morgan,[39] with the idea of providing students with a linguistic, cultural, and personal growth model such as that described by Carter and Long.[40] We also took into account the theories of neuroteaching[41] that recommend the use of art in teaching methodologies outside areas of art in the curriculum, in order to help create enduring learning and increase student motivation. These were the main objectives of the project:

- Make foreign language classes more dynamic through the use of literature and technology.
- Encourage collaborative work among students.
- Use literary resources (haiku) in the foreign language classroom in order to work on the oral and written skills of the language and to develop its linguistic competence.
- Develop positive values to nurture and increase environmental awareness through reading and writing haiku.
- Develop students' literary competence in a foreign language through poetry (haiku).
- Conduct poetry recitals in English with haikus after working on them in the classroom (because of the pandemic the recitals were performed through videoconferences).
- Exhibit the haiku in the form of photo-poems (due to the pandemic the exhibitions were virtual).
- Learn different collaborative methodologies and the use of virtual tools in order to make classroom activities more dynamic, with the aim of acquiring skills and teaching resources for both future teachers and educators.

The course contents that our students were required to learn to develop their communicative competence were mainly linguistic, including those related to grammar and basic vocabulary. The curricular level corresponded to an Intermediate/Upper-Intermediate level of English,[42] the equivalent of a B1-B2 level as established by the standard of the Common European Framework of Reference for Languages (CEFR),[43] updated in 2020 by the Council of Europe, the approximate equivalent to other global English evaluation schemes, such as Cambridge ESOL, Canadian Academic English Language Assessment (CAEL), or TOEFL.

"The Poetry Reading Project" endeavored not only to increase students' possibilities of learning by enriching classes with the literary resource (haiku), but also to develop creative and critical student abilities and competencies related to their sense of nature, and their environmental values. Our own motivation behind this methodological proposal is based largely on our belief that for mindful learning to occur it is necessary to go beyond textbooks and put language into real contexts of use. Moreover, we conceive the language teacher as a kind of cultural intermediary and guide and, as such, consider it necessary to approach language from a perspective where culture is present, given that it is an essential component in global education. In this sense, art, and specifically literature, as a maximum exponent of culture, can allow us to work on the emotional aspect, ensuring a more meaningful learning while seeking to awaken the ecological sensibility and environmental imagination of our students. More importantly though, if we ascribe to Orr's idea of ecological literacy, we must also acknowledge the role of the language teacher as a mediator between culture and environment.

For our project, the classes were then organized following a practical schedule drawing on pedagogical plans previously suggested in the form of a workshop, where students first read haiku and then created them with the help of technology webs and apps.[44] These virtual tools were used for educational purposes to enhance the activity but became essential during the online teaching period that followed in Spain and most of Europe as well as the rest of the continent after March 14, 2020; as a result of the pandemic, all instruction was forced to become, literally overnight, virtual.[45] Faced with this new educational scenario of online teaching we doubted that the project that we had proposed for the course could be carried out, due to the fact that we found ourselves immersed in a new way of virtual teaching, one which we were not familiar with and which eliminated real face-to-face contact. However, much to our surprise, not only did the digital tools and applications we used allow us to continue working along the lines of our project, but this new virtual scenario would also serve to develop the students' environmental competence by fostering their critical thinking, through haiku, while learning a new language.

The first task was to familiarize students with basic notions of poetry and help them become aware of the importance of rhythm, word stress, and the number of syllables. Following this, we selected a group of poems adapted to the students' level of language in order to be read and understood in class to facilitate the proposed tasks. In this sense, it is important to underline that the criteria for the selection of poems may vary depending on the objectives and the educational contents in each context. In our case, we used the following selections:

- "Windy Nights" by Robert Louis Stevenson
- "Colours" by Christina Rossetti
- "The Months of the Year" by Sara Coleridge
- "You Are Old, Father William" by Lewis Carroll
- "Morning Has Broken" by Eleanor Farjeon

The selection criteria were that poems were appropriate for the students' level, related to the educational content, written by classic authors in English, and accessible on the internet.

The dynamics of the class had two differentiated parts. First, students worked in groups to read the poem and answer questions related to its meaning as well as formal aspects of the poem, such as the number of verses and syllables of each verse. Then students worked individually on the creative writing proposal. At the end of each class session, both the group and individual work was shared and corrected with the collaboration of the rest of the class. Students then performed different creative tasks to put into practice the previous notions. These activities included the following:

- A poetic collage—a poem created by using the verses of the poems previously read;
- Blackout poetry—poetry created by crossing out the words of one of the poems previously read;
- A collaborative poem—a poem created by groups of students, where each student in the group writes a verse;
- A calligram—a creation of a calligram with one of the poems;
- Cyberpoetry—a creation of cyberpoetry using apps such as Poetry Creator,[46] Blackout Bard;[47]
- Twitter poetry—a creation of a poem, with the collage technique, using the tweets of famous politicians or celebrities who appeared in the social network Twitter.

To introduce students to haiku, we used the traditional Japanese haiku and format (number of syllables, nature and seasons as the central theme, the use of the five senses as a literary resource, and the use of present verb tenses).[48] In order to identify the most important aspects of haiku, we chose some of Matsuo Basho's most representative haikus from the book *Narrow Road to the Interior*,[49] which were read in their English translation. These included examples such as haiku titled "The Orchid's Perfume," "Spring Passes," or "The Ivy's Planted." Finally, after reading Basho's haiku, it was time for students to create their own with the aid of virtual tools. We hosted a cyberhaiku session in which students had to read and create haiku about topics related to the environment using apps in their mobile phones, such as the Interactive Haiku Generator[50] and the Poem Generator,[51] which can be used to create a haiku in seconds with the words selected.

However, the final activity to round off the classes was without doubt the most popular. This consisted of a virtual haiku exhibition, which was developed during the middle of March when virtual classes first began. The task was based on environmental haiku, written during the confinement by students, under the title "From My Window." Given that these students had already read haiku and were familiar with the format, they were asked to write a poem in the form of a haiku related to the theme of nature by describing what they could see from their window. They were also invited to accompany it with a photo. The poems were then shared by all the class and posted on a digital wall[52] in order to be visible for everyone. Given their reduced length and evocative imagery the creation of haiku proved to be a successful way to encourage creativity. Furthermore, these simple manifestations often served as an emotional outlet by allowing students to reflect on their own feelings during the pandemic and confinement, often portraying nature while connecting them to the natural world. This was followed at the end of the academic year by a virtual poetry slam, as a culmination of the activities carried out, so that students could perform their poems for each other.

With reference to the follow-up evaluation of the project, fifty-seven haiku were created. Out of these, 96.4% of the haiku (fifty-five poems) contained the correct number of verses, 57.8% (thirty-three poems) of the poems contained the correct traditional pattern of 5-7-5 syllabus, and 98.2% (fifty-six poems) of the total were related to nature. As the results indicate, the overall objectives were achieved by the students (poems followed the number of verses and were related to the theme of nature), although according to the data some students needed to work more on the syllable count in order to keep within the traditional pattern.

The following are only a few examples of the haiku students produced and shared in the virtual wall:[53]

The mountains rule all / with their vivacious greatness / crowned by the sunlight.

I see a sad sky / With many days without you / That just wants to cry.

The Birds chitchatting / With the flowers in full bloom / And the sound of the wind.

These poems, inspired by the prompt "From My Window," show from very different perspectives what many of the students may have taken for granted up until then: little pieces of nature that could be seen from their windows while being locked down at home. When our students looked out of their windows at the world they could no longer roam freely in, they saw an outside world where nature, with all its greenery and splendor, its birds, insects, and other creatures, stood apart from humanity as a kind of message of salvation. Writing haiku meant seeing the world with new eyes, observing nature with the hope of reconnecting our inner world, behind the window, to the natural world that continued to thrive outside. In this case we were forced to stand back and enquire into the many ways we take the world for granted, obliged to slow down and observe nature.

By writing haiku students were able to learn English while developing their sense of environmental awareness. This is becausewriting poems about aspects of the natural world will help students to focus their gaze on them, a gaze that, perhaps, is unfocused and distant from the natural world that surrounds us. In a sense, this may be due to the accelerated modern times or to life in the cities—to give some examples—styles of living that could produce in us a rupture with environmental values. Therefore, dealing with nature as a topic, and watching the seasons and writing haiku, became in some cases a song of hope. However, in other cases one could glimpse the loneliness felt and the despair, making Orr's words about ecological literacy ring true when he speaks about "the capacity to observe nature with insight"[54] or the idea of connecting human minds with nature as Stibbe[55] mentions. This was also accompanied by a catharsis of emotions in the Aristotelian sense (as Aristotle expressed in his *Poetics*), in so far that students were able to communicate the emotions of that unforgettable moment—which took place during the online classes held during the first wave of the COVID-19 pandemic—with the rest of their classmates in a more profound way. By involving their emotions in the learning process its impact became more effective.

Finally, given that the teacher is merely a guide, we consider it appropriate to include some of the feedback we received from students at the end of the course, reflecting their own feelings about the project and how they might envision using it in the future. Below are some of the most representative comments:

> This alternative gives us the opportunity to introduce poems in early childhood education, through different methodologies. In addition, in some activities they encourage the development of creativity and the acquisition of vocabulary in another language. From my point of view, it is quite interesting and curious to be able to work this technique with young children.

> Sometimes it frustrated me a bit because if it is already a bit difficult for me to write poetry in Spanish, doing it in a foreign language increases that difficulty. However, I believe that through the work of poetry we can come to learn English, not only in its standard form, but also in its figurative use, as well as playing with language, which is very enriching. There were also specific activities such as creating a poem with tweets on Twitter that I found extremely entertaining and in general I was happy with the result.

> It was a different way of facing the subject that I had not experienced up to now and I believe that it can have benefits in the classroom.

> I consider it to be a way to bring poetry and the English language closer to children. We always focus on studying and learning based on narrative texts, stories, stories, etc. But poems also provide us with a way of learning English that also allows us to approach a language where aesthetics predominate. It has seemed like a very interesting way to learn English.

> It seemed interesting to me, but I wish it had been more focused on Early Childhood Education.

As can be seen, overall, the students' opinion in relation to the project was positive; they viewed it as a creative proposal that as teachers they could develop for use in their classrooms someday. Regarding possible improvements for the project, as the comment of the last student suggests, it would perhaps be interesting to work specifically on children's poetry to make it more relevant and useful for their work as future educators.

CONCLUSIONS

Developing critical thinking about what is happening with regard to the preservation of nature and the future of our planet is crucial for future generations and should be of major concern not only to all educating governments and schools but to each of us as individual teachers and educators, school and community leaders. The nature of haiku can make this possible through its focus on the beauty of nature and the endurance of the moment.

The main purpose of reading and writing haiku is sharing moments of our lives and experiences that have moved us, such as living through a pandemic, confined and deprived of freedom of movement. At the deepest level, this is surely one of the greatest purposes of all art, not just poetry.

As teachers and educators, we believe it is important to be creative and innovative when planning our classes and developing an approach based on different ways of accessing literary texts, choosing one for close analysis, or organizing the syllabus from an eclectic model to favor required educational objectives. Introducing poetic and educational encounters and experiences such as those described can make the poetry reading and writing process an unforgettable educational experience in the classroom, while fomenting environmental awareness and, we hope, encouraging future educators and teachers to follow and explore these paths.

It is our hope that by nurturing a sense of wonder this knowledge of the natural world will carry forward into the future and affect the ecological well-being of our students and of future generations, moving beyond the present crisis toward a more sustainable way of living with Earth.

If, as Abram suggests, we have lost touch with the world around us then perhaps haiku can be a useful tool to bring us back, to reconnect with nature. It is in this sense that haiku might be said to be "the place where we save the earth."[56]

NOTES

1. See Z. Grange et al., "Ranking the Risk of Animal to Human Spillover for Newly Discovered Viruses." PNAS April 13, 2021. https://www.pnas.org/content/118/15/e2002324118

2. Val Plumwood, *Environmental Culture: The Ecological Crisis of Reason* (London: Routledge, 2002), 8.

3. See Peter Grundy, "Humanistic Language Teaching." In *Routledge Encyclopedia of Language Teaching and Learning*, edited by Michael Byram (London: Routledge, 2004), 282–285.

4. For more information see https://www.coe.int/en/web/north-south-centre/global-education.

5. Maastricht Declaration: European strategy framework for improving and increasing global education in Europe to the year 2015, resulting from the first Europe-wide global education congress organized by the North-South Centre of the Council of Europe, Maastricht, The Netherlands, November 15 to 17, 2002.

6. Kip Cates, "Global Education." In *Routledge Encyclopedia of Language Teaching and Learning*, edited by Michael Byram (London: Routledge, 2004), 241.

7. See James M. Bourke, "Designing a Topic-based Syllabus for Young Learners." *ELT Journal* 60, no. 3 (2006): 279–286.

8. Oxford Advanced American Dictionary https://www.oxfordlearnersdictionaries.com/definition/american_english/foreign.

9. See Carmen Muñoz, *Aprender idiomas* (Barcelona, Paidós, 2002).

10. See Ronald Carter and Michael Long, *Teaching Literature* (London: Longman, 1991); Alan Duff and Alan Maley, *Literature* (Oxford: Oxford University Press, 1990); Gillian Lazar, *Literature and Language Teaching* (New York: Cambridge University Press, 1993) and Tom Hunley. *Teaching Poetry Writing: A Five-Canon Approach* (Buffalo, NY: Multilingual Matters, 2007).

11. See Francisco Mora, *Neuroeducación: Solo se puede aprender aquello que se ama* (Madrid: Alianza Editorial, 2015).

12. Glen Whitman and Ian Kelleher, *NeuroTeach: Brain Science and the Future of Education* (Maryland: Rowman & Littlefield, 2016), 91.

13. Luke Rinne, Emma Gregory, Julia Yarnolinskaya, and Mariale Hardiman, "Why Arts Integration Improves Long-Term Retention of Content," *Mind, Brain, and Education* 5, no. 2 (2011): 89–96.

14. David A. Sousa, *How the Brain Learns* (Thousand Oaks: Corwin, A SAGE Company, 2011).

15. See Peter Chofer, "Literary Theory and Literature Teaching." In *Routledge Encyclopedia of Language Teaching and Learning*, edited by Michael Byram (London: Routledge, 2004): 397–393.

16. See María Encarnación Carrillo-García and Aurora Martínez-Ezquerro, "Neurodidáctica de la Lengua y la Literatura." *Revista Iberoamericana De Educación* 78, no. 1 (2018): 149–164.

17. David Hanauer. "Meaningful Literacy: Writing Poetry in the Language Classroom." *Language Teaching* 45, no. I (2012): 105–115.

18. David Orr, "Environmental Education and Ecological Literacy." *Education Digest* 55, no. 9 (1990): 49.

19. David Cobb, *Haiku* (New York: Overlook Press, 2013), 5.

20. Lucien Stryk, *On Love and Barley: Haiku of Basho* (Honolulu: University of Hawaii Press, 1985), 15.

21. David L. Barnhill, *Basho's Haiku: Selected Poems of Matsuo Basho* (Albany: State University of New York Press, 2004), 7.

22. In Janine Beichman, *Masaoka Shiki: His Life and Works* (Boston: Twayne Publishers, 1982), 31.

23. Yoel Hoffman, *Japanese Death Poems: Written by Zen Monks and Haiku Poets on the Verge of Death* (Rutland: Charles E. Tuttle Co, 1996), 39.

24. Carson's essay "Help Your Child to Wonder" first appeared as an article in the magazine *Woman's Home Companion* in July 1956 and was later published posthumously by Harper & Row as *The Sense of Wonder* in 1965.

25. See David Abram, *The Spell of the Sensuous* (New York: Pantheon Books, 1996).

26. Abram, 273.

27. Arran Stibbe, *Animals Erased: Discourse, Ecology and Reconnection with the Natural World* (Middleton: Wesleyan University Press, 2012), 162.

28. David Orr, *Ecological Literacy: Education and the Transition to a Postmodern World* (Albany: State University of New York Press, 1992), 86.

29. David Orr, *Earth in Mind: on Education, Environment, and the Human Prospect* (Washington, ML; London, UK: Island Press, 1994), 2.

30. Orr, *Ecological Literacy*, 91.

31. See Atsushi Iida, "Poetry Writing as Expressive Pedagogy in an EFL Context: Identifying Possible Assessment Tools for Haiku Poetry in EFL Freshman College Writing." *Assessing Writing* 13, no. 3 (2008): 171–179.

32. Atsushi Iida, "Expressing Study Abroad Experiences in Second Language Haiku Writing: Theoretical and Practical Implications for Teaching Haiku

Composition in Asian EFL Classrooms." *Asian English Language Classrooms* (2017): 180–191.

33. Maria Moss, "Writing Creatively in a Foreign Language: Vignettes, Haikus, and Poetry." *Zeitschrift für Interkulturellen Fremdsprachenunterricht* 25, no. 2 (2020): 29–53.

34. Yong B. Lee, "The Practice of Haiku Writing in Second Language Classrooms." *Komaba Journal of English Education* 2 (2011): 23–44.

35. Merrill Swain, "Three Functions of Output in Second Language Learning." In *Principle & Practice in Applied Linguistics*, edited by G. Cook and B. Seidlhofer (Oxford: Oxford University Press, 1995), 125–144.

36. The project "The Poetry Reading Project" was an innovation project granted by the Innovation Unit of the University of Murcia and was developed during the 2019 to 2020 academic year by Dr. María Encarnación Carrillo-García and Dr. Lorraine Kerslake. We want to thank the students from the University of Alicante and the University of Murcia for their enthusiastic participation in this project and to acknowledge their permission to use the poems, data, and opinions collected in this article.

37. The majority of students were language learners with an Advanced level of English, corresponding to a C1 level of English in the European Framework.

38. https://www.um.es/web/innovacion/.

39. See Carol Morgan, "Poetry" in *Routledge Encyclopedia of Language Teaching and Learning*, edited by Michael Byram (London: Routledge, 2004), 470–472.

40. Ronald Carter and Michael Long, 1991.

41. See Glen Whitman and Ian Kelleher, *NeuroTeach: Brain Science and the Future of Education* (Maryland: Rowman & Littlefield, 2016).

42. In relation to the learning standards, these refer to the curricular levels that language students must reach, levels established for the European Union as a whole, and which were also considered as part of the objectives for this project.

43. See The Council of Europe. Common European Framework of Reference for Languages: Learning, Teaching, Assessment (Updates the CEFR 2001). Council of Europe, 2020. https://www.coe.int/en/web/common-european-framework-reference-languages.

44. https://aulavirtual.um.es/portal, https://zoom.us, https://padlet.com.

45. It is important to recall that Spain suffered one of the most severe and strict lockdowns throughout the country with home confinement and suspension of all non-essential business activity imposed throughout as of March 14, 2020. In education this meant that classes became virtual for the rest of the term. In September schools and universities reopened fully with new rules introduced. These included that all students over six had to wear masks, class sizes were reduced, students were separated into assigned "bubbles" to prevent mixing, and desks were kept 1.5 meter apart.

46. apps.apple.com/us/app/poetry-creator-verses-poetry-poems-poets/id371925480.

47. play.google.com/store/apps/details?id=blackout.one3one4.com.blackout&hl=en_US.

48. Particularly instructive is *How to make a Haiku* by the JAL Foundation (2014). For more information see: www.jal-foundation.or.jp/contest-e.html.

49. See Matsuo Basho, *Narrow Road to the Interior and Other Writings* (Boston: Shambhala, 2012).
50. www.languageisavirus.com/interactive-haiku-generator.php?ac=Generate#.X-RtvC3FRYI
51. www.poem-generator.org.uk/haiku/.
52. es.padlet.com/elauladeingleseninfantil/z86qny96vfns.
53. es.padlet.com/elauladeingleseninfantil/z86qny96vfns.
54. Orr, *Ecological Literacy*, 86.
55. Arran Stibbe, *Animals Erased: Discourse, Ecology and Reconnection with the Natural World* (Middleton: Wesleyan University Press, 2012), 162.
56. Jonathan Bate, *The Song of the Earth* (Cambridge: Harvard University Press, 2000), 283.

REFERENCES

Abram, David. 1996. *The Spell of the Sensuous*. New York: Pantheon Books.

Aristotle. 2008. *Poetics*. Translated by S.H. Butcher. The Project Gutenberg Ebook: www.gutenberg.org/files/1974/1974-h/1974-h.htm.

Barnhill, David L. 2004. *Basho's Haiku: Selected Poems of Matsuo Basho*. Albany, NY: State University of New York Press.

Basho, Matsuo. 2012. *Narrow Road to the Interior and other writings*. Boston: Shambhala.

Bate, Jonathan. 2000. *The Song of the Earth*. Cambridge: Harvard University Press.

Beichman, Janine. 1982. *Masaoka Shiki: His Life and Works*. Boston: Twayne Publishers.

Bolt Simons, Lisa M. 2015. *Haikus*. USA: The Child's World.

Bourke, James M. 2006. "Designing a Topic-based Syllabus for Young Learners". *ELT Journal* 60(3): 279–286.

Carter, Ronald and Michael Long. 1991. *Teaching Literature*. London: Longman.

Carson, Rachel. 1965. *The Sense of Wonder*. New York: Harper & Row.

Carrillo-García, María Encarnación, and Aurora Martínez-Ezquerro. 2018. "Neurodidáctica de la Lengua y la Literatura." *Revista Iberoamericana De Educación* 78(1): 149–164.

Cates, Kip A. 2004. "Global Education." In *Routledge Encyclopedia of Language Teaching and Learning*, edited by Michael Byram. London: Routledge: 241–242.

Chofer, Peter. 2004. "Literary Theory and Literature Teaching." In *Routledge Encyclopedia of Language Teaching and Learning*, edited by Michael Byram. London: Routledge: 397–393.

Cobb, David. 2013. *Haiku*. New York: Overlook Press.

Duff, Alan and Alan Maley. 1990. *Literature*. Oxford: Oxford University Press.

Grundy, Peter. 2004. "Humanistic Language Teaching." In *Routledge Encyclopedia of Language Teaching and Learning*, edited by Michael Byram. London: Routledge: 282–285.

Hanauer, David. 2012. "Meaningful Literacy: Writing Poetry in the Language Classroom." *Language Teaching* 45(I): 105–115.

Hoffman, Yoel. 1996. *Japanese Death Poems: Written by Zen Monks and Haiku Poets on the Verge of Death.* Rutland, VT: Charles E. Tuttle Co.

Hunley, Tom. 2007. *Teaching Poetry Writing: A Five-Canon Approach.* Buffalo, NY: Multilingual Matters.

Iida, Atsushi. 2008. "Poetry Writing as Expressive Pedagogy in an EFL Context: Identifying Possible Assessment Tools for Haiku Poetry in EFL Freshman College Writing." *Assessing Writing* 13(3): 171–179.

Iida, Atsushi. 2017. "Expressing Study Abroad Experiences in Second Language Haiku Writing: Theoretical and Practical Implications for Teaching Haiku Composition in Asian EFL Classrooms." *Asian English Language Classrooms*: 180–191. https://doi.org/10.4324/9781315755243.

Jal Foundation. 2014. *How to Make Haiku (English).* Tokyo: Jal Foundation.

Lazar, Gillian. 1993. *Literature and Language Teaching.* New York: Cambridge University Press.

Lee, B. Yong. 2011. "The Practice of Haiku Writing in Second Language Classrooms". *Komaba Journal of English Education* 2: 23–44.

Mora, Francisco. 2015. *Neuroeducación: Solo se puede aprender aquello que se ama.* Madrid: Alianza Editorial.

Morgan, Carol. 2004. "Poetry," in *Routledge Encyclopedia of Language Teaching and Learning*, edited by Michael Byram. London: Routledge, 470–472.

Moss, Maria. 2020. "Writing Creatively in a Foreign Language: Vignettes, Haikus, and Poetry." *Zeitschrift für Interkulturellen Fremdsprachenunterricht* 25(2): 29–53.

Muñoz, C. 2002. *Aprender idiomas.* Barcelona: Paidós.

Fidalgo, María Virtudes Núñez. 2020. "La literatura en el marco de las neurociencias cognitivas: Nuevas perspectivas de estudio." *Archivum: Revista de la Facultad de Filología* 70: 165–191.

Orr, David. 1992. *Ecological Literacy: Education and the Transition to a Postmodern World.* Albany: State University of New York Press.

Orr, David. 1990. "Environmental Education and Ecological Literacy." *Education Digest* 55(9): 49–53.

Orr, David. 1994. *Earth in Mind: on Education, Environment, and the Human Prospect.* Washington, ML; London, UK: Island Press.

Plumwood, Val. 2002. *Environmental Culture: The Ecological Crisis of Reason.* London: Routledge.

Rinne, Luke, Emma Gregory, Julia Yarnolinskaya, and Mariale Hardiman. 2011. "Why Arts Integration Improves Long-Term Retention of Content." *Mind, Brain, and Education* 5(2): 89–96.

Sousa, David, A. 2011. *How the Brain Learns.* Thousand Oaks: Corwin, A SAGE Company.

Stibbe, Arran. 2012. *Animals Erased: Discourse, Ecology and Reconnection with the Natural World.* Middleton: Wesleyan University Press.

Stryk, Lucien. 1985. *On Love and Barley: Haiku of Basho.* Honolulu: University of Hawaii Press.

Swain, Merrill. 1995. "Three Functions of Output in Second Language Learning," in *Principle & Practice in Applied Linguistics*, edited by G. Cook and B. Seidlhofer. Oxford: Oxford University Press, 125–144.

Whitman, Glen and Ian Kelleher. 2016. *NeuroTeach: Brain Science and the Future of Education*. Maryland: Rowman & Littlefield.

Chapter 8

Introducing Sustainability Topics with Ursula Le Guin's "The Ones Who Walk Away from Omelas" and Richard Powers's "The Seventh Event"

Sarah Wyman and Rachel Cohen

In this chapter, we define ecocriticism as a discipline that ties the study of literature to environmental/social/economic sustainability projects. Our goal is student empowerment through an understanding of the way literature and the arts can both reflect and impact social change. Storytelling in words and images introduces and disseminates new ideas through its themes and arguments; moreover, it acts as a commons-of-thought or an arena of invention where new conceptualizations and innovations come to be. Students are introduced to a range of sustainability frameworks through which to study two celebrated stories of contemporary U.S. literature. They build intercultural competency by reading about alternate ways of walking through the world; they consider their own views and values by unpacking the real and imagined dilemmas these stories present; they balance individual accountability against the power of collective action; and through their vicarious experience of characters' comments and choices, they imagine themselves into another's perspective. As Martha C. Nussbaum argues, striving to see the world through the eyes of another develops empathy and is a fundamental affordance of the liberal arts education, essential to building a successful society.[1] Students may do this through active engagement with the course content, with their instructors, and with their peers. An introduction to the notion of systems thinking helps illuminate the features of intersectional sustainability schemas (environmental, social, and economic) that describe the complex problems threatening our planetary and species survival.

Ursula Le Guin's "The Ones Who Walk Away from Omelas" (1973)[2] presents an "allegory of the American Consciousness," in the author's words, that corresponds to real-world systems that perpetuate exploitation and other forms of inequality. Using the utilitarian philosophy of pragmatist William James, she explores the limits of the social contract and of humanity's willingness to accept and normalize the suffering of others. Le Guin actively engages the reader in a position of responsibility when she employs second-person narration. Richard Powers's short story about storytelling, "The Seventh Event" (2005),[3] investigates the intersection between science and literature, two representational modes that describe and define the world. The author's fictional character Mia Erdmann desires to create life in the face of mass extinction, first through bioengineering, and then through the act of producing fictions. Using reverse chronological order and a highly fragmented story line full of intertextual references, Powers creates an unconventional narrative structure that makes sense once the underlying systems of thought become clear. Not only are science and fiction contrasted, but a sliding scale from the atomic to the cosmic level expresses humans' relative insignificance as well as the urgency of their active engagement in changing the environmentally lethal status quo. Each of these stories can be read in light of the intersecting environmental, social, and economic impacts of climate change and of a world order that fails to deliver adequate opportunities, rights, and protections to all.

These approaches provide training in literary analysis and interpretation for advanced high school students or undergraduate and graduate students in college/university, as the goals are to raise sustainability awareness and understanding, practice critical thinking, develop information literacy, and empower students to be informed and empathetic change agents (all transferable skills for professional success). Students study conceptually complex stories and are assessed on their reading comprehension and written expression. They practice using rhetorical terminology in order to speak with precision about the use of language with special attention to point of view, plot structure, narrative voice, intertextual reference, and allegorical features. Students practice oral expression in structured discussion or low-stakes presentation activities, and they each write a formal argumentative essay in which they identify and define a dilemma within one of the short stories or possibly, other recommended readings. Students encounter a variety of sustainability frameworks that endorse systems thinking, international collaboration, potential solutions to climate change, and environmental justice as it relates to diversity, equity, and inclusion. A study of views and voices within the short works of fiction connects to cultural perceptions of social inequality and the disproportionate burden of climate-based threats borne by the underrepresented and disenfranchised.

This course module introduces students to the concept of ecocriticism as a theoretical approach to reading literature and to sustainability frameworks that may include the *Hau de no sau nee Address to the Western World* (1977), the sustainability definition from The Brundtland Commission, *Our Common Future* (1987); The Principles of Environmental Justice (1991); The Earth Charter (2000); Oxfam's Donut Model (2012) for a "safe and just space for humanity"; the United Nations Sustainable Development Goals (SDGs) (2016); Julian Agyeman's Just Sustainabilities (2003); and Project Drawdown climate solutions. By looking at the United Nations SDGs set of seventeen categories as a figurative map of the world and a manifesto, students will have the opportunity to critique a model of international collaboration developed by 193 countries.

An ecocritical lens allows students and instructors to distinguish between deep ecology and social ecology as they connect Le Guin's and Powers's short fictions to the project of sustainability education and action. An ecofeminist lens introduces a consideration of gender dynamics and an ethics of care[4] that privileges communication and collaboration. Using the concept of systems thinking, students approach the global project of sustainability outlined by the United Nations SDGs and additional sustainability frameworks through an intersectional lens that emphasizes the interrelation of environmental, social, and economic factors and impacts.

By acknowledging challenges, promising possibilities for learning and growth, and providing the guiding principles of sustainability frameworks, educators can help students obtain the tools and build the skills they need to contribute to the global project of sustainability. This involves fostering the acquisition of technical skills, historical knowledge, and interpersonal competency. As Cornell University professor Neil Lewis has expressed, we need to send intentional messages about how people might behave toward greater sustainability by building partnerships, working across difference, encouraging empathy as a sustainability goal, and fostering intercultural competency in our communities, on our campuses, and as global citizens.[5] The objective of this curriculum, to borrow goals articulated by CUNY educator Bret Eynon, is to foster resilient problem-solvers who are agile, integrative thinkers who value collaborative practices and become lifelong learners.[6] The information is out there; the educator's role is to shape the learning experience and steward the student through a process of creating new knowledge and self-understanding. To reference Bloom's *Taxonomy of Learning Outcomes*, students should develop skills to evaluate, to decide; to analyze, to connect; to apply, to use; to comprehend, to explain; and to remember and think critically about course content and their formative experiences.[7] Key student learning outcomes and experiences are highlighted in the accompanying lesson resource for this chapter.

READING LE GUIN'S "THE ONES WHO WALK AWAY FROM OMELAS"[8] AND POWERS'S "THE SEVENTH EVENT" AS LITERARY MODELS FOR SYSTEMS THINKING ABOUT SUSTAINABILITY TOPICS

This section provides an introductory reading guide for the instructor for Ursula Le Guin's "The Ones Who Walk Away from Omelas" and Richard Power's "The Seventh Event." After each synopsis and discussion, key sustainability themes are highlighted as well as the literary considerations that factor into the analysis and interpretation learning outcomes.

Ursula Le Guin, "The Ones Who Walk Away from Omelas"

In "The Ones Who Walk Away from Omelas (Variations on a Theme by William James)," Ursula Le Guin presents a utopia that turns out to include an imperfect, even nightmarish dystopia. The tension between these two heaven-and-hell extremes could be summed up in a pull between the impulse to leave this perfect world, an action depicted in the title, and the joyous arrival of the festival that sets the stage: a carefree community that seems pleasing and just turns out to be structured on injustice and inequality, making it ultimately unsustainable for some of its citizens. Ethical confusion arises both within the fictional world of the story and when the reader attempts to reconcile that textual space with the real-world counterpart to which it refers. Le Guin considers the story an allegory of U.S. culture at the time of the Vietnam War, inspired by the "shock of recognition" she experienced upon reading this passage from William James:

> If the hypothesis were offered us of a world in which . . . millions [should be] kept permanently happy on the one simple condition that a certain lost soul on the far-off edge of things should lead a life of lonely torment . . . even though an impulse arose within us to clutch at the happiness so offered, how hideous a thing would be its enjoyment when deliberately accepted as the fruit of such a bargain?[9]

Le Guin contends that in this passage, "the dilemma of the American conscience can hardly be better stated."[10] By extension, Le Guin's famous story may be read as an allegory of the plight of the planet and the ultimate costs of a profit-driven global economy of extraction and consumption. Or, the tortured child in the basement, undefined by gender, race, class, or nationality, may come to represent the disenfranchised members of humanity who bear a disproportionate burden caused by the effects of climate change. Fire, famine, and flood, for example, may be natural disasters, but still demonstrate environmental injustice, economic exploitation, and other forms of social inequality when unchecked by policies that might ensure equity and adequate protections for their victims.

In the story, a community inhabits a wonderful, fairy-tale world free of illness, anxiety, and social strife. The Omalasians subscribe to a social contract contingent upon the exploitation of one child, trapped and deprived in a dark basement beneath the city's civic buildings, to secure the happiness of all other citizens. The author pits brute reality against fiction's capacity to conjure illusory solutions and offers no easy answers. Some mistakenly argue that Le Guin supports the noncompliance of those who "walk away" from an unbearable social contract. Neither Le Guin nor James, however, would necessarily applaud the members who choose to leave this community, for to do so would not change the status of the suffering child. While the choice to walk away from the "hideous bargain" Le Guin puts forth may seem correct at first, a more careful reading suggests that both Le Guin and James would elect to stay in Omelas, imperfect as it turns out to be. Each would likely insist on a dynamic ethical system existing and evolving among and dependent upon all community members. To withdraw from this fellowship would be comparable to betraying the social contract and abdicating responsibility for the child's lot; to leave would be to act individualistically, not in partnership with the collective. Rather than offer a pragmatic, utilitarian excuse—the good of the many outweighs the good of the one—Le Guin and James would place emphasis on the incarcerated "lost soul," who stands simultaneously inside and outside the society, sequestered in its dungeon in the center of town. Instead of dallying with escapist fantasies, the author engages the utopian trope in order to put forth a political statement that becomes increasingly clear as the story evolves.

A key aspect of this tale and a clue to Le Guin's opinion on the dilemma of departure manifests in the way the fiction itself contrives to hold the reader accountable as a virtual Omalasian, fully aware of parallels between the imaginary land and contemporary Western capitalist, imperialist society. The reader cannot not participate in the fiction, just as each citizen cannot *not* participate in Omelas' horrible social contract. Le Guin creates a narrator who invites the reader to wrestle with the moral dilemma presented while roping him/her/them into a terribly uncomfortable position by using several rhetorical tactics. The author sets up stylistic and structural tensions in order to lure the reader into making a choice that she refuses to make overtly. By involving the reader as co-creator of this fantasy world, she implicates the reader in the decision acted out by the characters, both those who stay and those who leave. "Perhaps it would be best if you imagined it as your own fancy bids," the narrator directs the reader, "for certainly I cannot suit you all."[11] So, Omelas becomes a relative utopia, perfectly modeled in the mind of each now-committed reader. The lushness of the language and the rhetorical power of the telling augment the seduction of this collaborative relationship between narrator and reader.

Le Guin starts us off in a highly poetic mode. She foregrounds the neat tension that will be the central dilemma of the story when she titles it with a "walk[ing] from," then starts the narration with a "[coming] to." Who will stay and who will leave? The main problem of the tale has already evidenced itself. She has stylized her text so that we are far from the realm of typical language of communication. Instead, our eyes and minds rest on the page where she foregrounds the materiality of language itself. The breathless, long lines of the first paragraph mimic the festival parade the words describe. Le Guin alliterates the sounds of the words, "mauve and grey, grave . . .,"[12] echoing the sonic features with a chiastic au-gr/gr-a flip. She employs the assonance of the long i-sound in the line, "high calls rising like the swallows' crossing flights . . . ," as the birds spread their wings.[13] The horses themselves are personified as the scene springs to life.

Next, the narrator makes two moves that jerk us out of traditional fairy-tale mode, waking us from the lull of gorgeous language. First, she takes a meta-textual (or meta-discursive) turn in which the text calls attention to its own project: telling a story. The narrative voice seems to stand outside the text and comment on it, reflexively. She problematizes her own tale-telling goal by questioning the effort of expression itself: "How is one to tell about joy? How describe the citizens of Omelas? . . . I wish I could describe it better."[14] This clumsy gesture comes across as ironic, following on the heels of the first paragraph's masterful display of technical skill in crafting a story and engineering effects of sound. The narrator's move to divest herself of authority and responsibility for defining Omelas further implicates the reader in completing the construction of what turns out to be an anti-utopia, after all.

In a second important turn, the narrator addresses the reader directly. Thus, we are drawn into the story, as virtual or contingent inhabitants of Omelas. Once the narrator starts generalizing about our "bad habit," the reader becomes a bit uncomfortable, without a voice of her own in the text with which to explain or defend herself. We do not want to accept the narrator's definition of us: "We have a bad habit . . . of considering happiness as something rather stupid."[15] Echoing the philosophy of Hannah Arendt, the narrator dismantles the very artistry she just demonstrated, "This is the treason of the artist: a refusal to admit the banality of evil and the terrible boredom of pain."[16] We begin to understand that our delicious picture of the city of Omelas actually disguises a dirty secret, necessary to its survival. We eagerly let ourselves be drawn into the initial festivities, only to find ourselves abruptly confronted with a serious moral dilemma.

The ungendered, incarcerated child locked in a basement serves as a trope of suffering, a ritual sacrifice, and bears a horrifying burden for the sake of its fellow citizens' joy. The child held captive under unspeakable circumstances in an ambiguously public/private space functions as a scapegoat for the town. Le Guin reports that she imagined this city after driving away from Salem, Oregon, seeing the town's name reflected backward in her rearview mirror. Through this linguistic reversal, she evokes the notorious colonial New England town where fourteen women deemed witches were executed to preserve the "public good" between 1692 and 1693. Thus, an historical allusion sets the stage for a high-stakes morality tale of invaluable sacrifice.

Neither choice, to stay or to leave, is without serious consequences, for even those who leave have not escaped responsibility for what goes on in the basement. To refuse to participate in this outrage by partaking in the happiness it potentially engenders does not solve the material situation of the suffering child. A citizen's departure ensures neither the child's comfort nor release. To leave is ironically, to not vote, to not act, to not work against climate change. In fact, several critics have pointed out the Omalasian's bad faith suggested by the weak defense of the child's incarceration, such as, "It is too degraded and imbecile to know any real joy."[17] The citizens attempt to justify their collective, callous torture of the child by stating that it is damaged beyond redemption, despite the fact that it has memories of an alternate existence and assumes it is being punished for misbehavior. Granted, more than one individual has left Omelas, so a new community could conceivably establish itself elsewhere. Le Guin refers to the "Ones," stressing their plurality and individuality at once. Yet, why should this uncharted realm that may not exist be any better at all? One can simply insist on the negative space of not-Omelas (the anti-anti-utopia) as symbolic of the community's betrayal and the danger of elevating individual choice above collective action and partnerships.

Utopias of any stripe exist as fixed fictions, and can only "seem" in a static, unchanging state, not "be." Yet, they can reflect or express phenomena parallel to real-world situations and potentialities. Le Guin's anti-utopia seems, by its very imperfection, to approach our world closely. The highly aestheticized nature of her story, its overt constructedness, should alert the reader to the deceptive enticements of art, the seduction of beauty, the distance between fiction and reality. The temptation to locate the story's treachery in the fiction itself rather than in the real world to which it corresponds parallels the mistaken characters' urge to leave Omelas altogether. While the impulse to flee this tainted paradise, this fictive realm may be revolutionary in itself, the act of going, when read against James's pragmatic philosophy, the traces of Le Guin's creative act, and their common belief in communal ethics, cannot be considered the morally sound or sustainable choice.

Richard Power's "The Seventh Event"

In "The Seventh Event," Richard Powers employs a framework that presents the inevitability of mass extinction, the "event" to which the title refers, through reverse chronological narration that narrows in scope from the annihilation of all species to the fictional death of a single character. Accordingly, the story is fractured into six sections, numbered backward, as the narrator recalls miscellaneous interactions and conversations on environmental sustainability and theories of replicating life with Mia Erdmann (translated as "my earthperson"), a fictitious colleague. Mia attempts to replicate or "create" life as a means to confront the threat of impending extinction, first through bioengineering and later, through writing fiction, which she comes to see as a more useful way of reproducing reality. Through these interactions, Powers explores the intersection between science and literature as representational modes that describe and define reality. By presenting environmentalist themes of species extinction and other threats to the planet linked to climate change, Powers ultimately demonstrates the need for a more systems-based approach to how humanity relates to the world at large.

In the opening sixth section, Powers addresses the highly technical perspective many eco-critics and environmentalists criticize, according to Ursula Heise, for "its presumption to know the natural world scientifically, to manipulate it technologically and exploit it economically, and thereby ultimately to create a human sphere apart from it."[18] Applying scientific and technological truths to the natural world, Heise argues, reduces nature to a "material resource and commodity."[19] As a result, humanity is alienated from its environment; nature becomes an artificial image or idea, and humans' limited capacity for perceiving the enormous dimensions of reality manifests as ambivalence toward that environment. The natural world is therefore, perceived in more conceptual terms, no longer rooted in a physical reality, and this creates a disconnect between body and place that perpetuates the destruction of natural resources for economic gain. As Mia defends her turn to a militant form of social ecology, she protests against a reverence for deep ecology or an (often romanticized) notion of nature as intrinsically beautiful and good: "Nature is not our zoo. We must get away from golly-green wide-eyed wonder. Wonder is too easy a dodge, a massive distraction from the more prosaic question of just what toxins we are sending downstream."[20] Nature boxed and sold as entertainment is one instance of humanity's artificial relationship with the environment; the consequences are planetary peril.

The most striking example of the damaging potential science proposes by separating human and nonhuman animals from the physical world is revealed through the narrator's memory of Mia in their college Chemistry class: "I remember our being assigned to determine the molecular weight of one particular unknown sample. We had a long flow chart of measurements and assays to perform."[21] Molecular weights of known substances, as opposed to unknown substances, are readily available for exploitation, and Mia takes advantage of this information:

> Forget that noise. Look at it. Feel it. Smell it. It's mashed up mothballs. Naphthalene. We know the molecular weight of Naphthalene. Save ourselves some work." Science, she declared, had to know when and how to cheat if it hoped to accomplish anything.[22]

Heise's concerns with applying numbers to the environment, and how this approach influences the meanings derived from the natural world, is evident in Mia's strategy to complete their chemistry experiment. Through Mia's inversion of the scientific method, starting with a conclusion rather than a hypothesis, the whole experimental procedure becomes obsolete under the known set of values or "truths" associated with Naphthalene.

The narrator comments on Mia's early bioengineering career, as it seems to guide her scrutiny of literary criticism (her later career). Academic literary analysis is held accountable for its failure to address the elaborate, contextual network intertwined with perceptions of the natural world and discursive traditions that create realities they claim to represent: "Literary analysis, she insisted, rightly explores the links between written worlds and the worlds they represent. But it wrongly stops, when making those links, at the gauge of the human."[23] This deficiency in literature that centers on the human and perpetuates illusory realities suggests, to Mia's disappointment, a potential, species-wide inability to perceive the "true scale of the world." Powers not only affirms this human incapacity via omission of a seventh section (the mass extinction event), but also by calling attention to Mia's own failure to communicate large-scale issues concerning species extinction through either numbers or fictions so as to effect change on a global scale. In the late stages of disease, as Mia's body shuts down, she thinks her last thoughts and relays a series of borrowed words to the narrator that become the titles for his story's six sections, communicating again, through already-existing structures of meaning.

Just as Powers's seemingly disjointed story coalesces around episodes in a lifelong friendship, the systems-thinking approach to environmental preservation both establishes order and works toward greater chaos. John Muir's quote begins the ironically final section one: "When we try to pick out anything by itself, we find it hitched to everything else in the Universe."[24] Heise states, however, that the early prototypes of an ideal ecosystem, concerned with "stability" and "harmony of all living things," which served as a blueprint for the "ideal model" of human habitation for the last century, are no longer supported by modern ecology.[25] Similarly, in her lab report, Mia asserts a false reality through a series of calculations that are mathematically valid, but only because they have been specifically manipulated to represent a procedure that was never actually performed. Mia's use of this shortcut demonstrates a concern for the "bigger picture"—the broadened perspective of human existence she aggressively seeks, but believes humanity is too self-oriented to fully grasp. Her career in bioengineering as a means to "make a living thing" thus demonstrates how a strictly statistical, scientific understanding of the environment is inadequate to conceptualize the intricacies of life on earth.[26] Although separation of the human from its environment underlies both perspectives, the former employs a more exploitative dynamic in its very disregard for the "little" in favor of the "big."

The theme of science as a representational mode of perceiving reality centered on humans extends throughout Mia's early bioengineering pursuits as well as her initial publication demanding "a literature that recalls the obscene majority of existence typically brushed aside by the novel of character revelation."[27] She writes, "If we could stop using nature as a metaphor for reflecting the human condition, we might begin to see how the human condition reflects some small fraction of nature's relentless urge to speculate."[28] This quote not only reflects ecocritical concerns with an anthropocentric or human-centered worldview, but also evokes the idea of life as "botched self-replication," a recycling phenomenon that applies to both literature and life itself on a molecular level. Her overarching argument asserts that nothing is original; humanity is narcissistic for seeing itself as separate from the environment, and that the use of nature as a metaphor perpetuates the very ignorance she criticizes. And yet, she cannot help herself, using anthropomorphic imagery and rhetorical contrast to make her points: "There are more cells in a baby's finger than people in the world."[29]

The irony of Mia's militant obsession with rejecting notions of "wonder," and all forms of literature that evoke it, becomes increasingly evident as the narrative scope decreases in accordance with Mia's shift in approach to replicating life. In maintaining an "objective" view of the world through science, she alienates herself from the rest of humanity, and in doing so, exhibits the same lack of self-awareness that she accuses both her critics and the literary canon of embodying. The narrator recalls,

> I asked her why not keep to the bigger story, if it's about to erase all others. She sighed, a little disgusted. "Because humans can't follow a bigger story. They don't want to read anything larger than autobiography. At least environmental justice squeezes the story down into a shape some people might recognize.[30]

As Mia nears the end of her struggle with Sandon's disease, a degenerative condition invented by Powers for the purpose of this story, she recognizes her own inability to dismiss feelings of "dread" as a symptom of depleted serotonin levels. One might expect the unconventional Mia to rationally welcome her own extinction, according to her cynical rant against "owl worshipper" environmentalists: "They think that advocating on behalf of some gigantic, fluffy species exonerates them. If they knew for one instant the pressure their own mere existence puts on the rest of the cracking web, they would take their own lives."[31] As she struggles, nevertheless, against her impending death, Mia speculates on an additional text she might publish before finally succumbing to her illness.

Mia's perceptual transformation grows increasingly apparent as the narrative continues to decrease in scope. Unlike sections five and six, which focus on Mia's attempts to replicate life through science, the conclusive section One consists of a document written by Mia herself, that characterizes fiction as the most practical means of achieving her goal of creation. The first page is a series of the six quotes used to introduce each of the six sections that collectively structure "The Seventh Event." The narrator's decision to incorporate these borrowed lines into his narrative reaffirms the notion of replication as a natural—and necessary—process by which creation can occur. The second page, however, is Mia's letter to the narrator containing a proposition for a story that he might one day choose to write. This prompt derives from an already-existing work of fiction that features two men bound to a hospital room due to their respective conditions. A heart patient, who at first occupies the bed beside the window, spends his days describing the scenery beyond the hospital walls to a person who is quadriplegic. In turn, the person with quadriplegia grows envious of the alleged view, and purposely knocks over a vial of vital medication when his roommate suffers a fatal heart attack. Once promoted to the bed beside the window, the partially paralyzed person discovers that the only perceptible sight is a "solid, grey wall" rather than the "constant circus of activity" advertised by the imaginative heart patient prior to his death.[32] Mia's proposition, which introduces a third patient, perpetuates the previous dynamic between the heart patient and his initial roommate. Relocated to the bed beside the window, the person with quadriplegia assumes the heart patient's role as storyteller, and the stroke victim, who occupies his former bed without a window, becomes the blind listener:

> There are three people in the room: the heart patient, the quadriplegic, and a stroke victim. And when the quadriplegic gets promoted to the window bed, and turns to see that solid grey wall, and the truth of what he's done crushes him, he stops, composes himself ... and tells the stroke victim: "You won't believe what's out there. You simply won't believe what I can see."[33]

Upon realizing that the view is not, in fact, anything but a blank wall, the character with a quadriplegic disability, like the heart patient before him, confronts his bleak reality by constructing fictitious descriptions of the outside world. This story is wholly dependent on the narrator's version of the story and thus exemplifies the significance of replication as a means of creating both life and literature. Without the narrator's "original" edition, which he himself stole from "some 1930's story omnibus that belonged to [his] father," Mia's subsequent version could not have been conceived in the first place.[34]

Not only does the premise of Mia's final story build upon a story that has already been written, but the proposition itself, in epistolary format, exists as a work of fiction within a larger work of fiction. This scale of a single death and discovery is grossly disproportionate to the unwritten "seventh" section that represents the vast, conceptual extinction of the human race, as well as Mia's condemnation of fiction as a whole. Back in section five, Mia writes, "And fiction, for its part, had for two centuries wallowed in a bottomless vanity that promoted the individual self to be the measure of all things: 'We have let our shot at awareness be bought off with a bauble.'"[35] Even though she initially criticizes her contemporaries for indulging in anthropocentric narcissism, she ultimately understands that what she once perceived as self-obsession is an inescapable facet of human experience borne of curiosity and the inescapable "wonder" she cannot ditch. The contrast between her early idealization of science and later appreciation of fiction is, therefore, representative of the narrow lens through which the broad network of human existence must be realized in order to confront the threat of climate change and human impact on the natural world. Furthermore, the inclusion of a fictitious text within the final section of "The Seventh Event" reinforces the notion of literature as, possibly, a more effective representational mode to confront environmental threats than science.

It is clear, by the end of section one, that the narrative itself develops along a trajectory that decreases in scope as the narrative progresses. While Mia comes to acknowledge humanity's limitations, including her own, paralyzed by disease, she understands the value of narrowing the elaborate network into a smaller frame, such as activism, as the most efficient way to communicate on the same level as her critics. Most importantly, however, Mia realizes that writing fiction is the closest that she, and humanity as a whole, will ever come to "making a living thing from scratch."[36] Even though this method of creation does not manifest in physical form, the final product, consisting of long-standing literary tropes and traditions, will exist as something "new" in theory. Similar to nature, which "tries everything at least once," literature is produced through copies of copies rather than spawned from original material, and Mia, on her deathbed, cannot surrender that energetic sense of wonder she once condemned in others.[37]

Richard Powers, through his own literary creations, thus urges his readers to acknowledge the detrimental, long-term impacts of mass delusion with the nonhuman world as well as the present, indisputably urgent necessity for radical sustainability initiatives across the globe. By characterizing Mia as a fictitious construct who strives to create life and functions as the subject of a larger work of fiction, Powers both demonstrates and imposes patterns of thought that underlie human perceptions—and by extension, treatment—of the natural world as the narrative progresses. As the framework shifts from an abstraction (the "seventh event" or species extinction) to fiction, Powers successfully creates a clear contrast between science and literature, and in doing so emphasizes the latter, particularly fiction, as the most effective medium through which humanity might examine the vast network of human experience if it hopes to save future generations from the ever-increasing threat of climate change and environmental destruction.

DISCUSSION

In the following section, we provide suggestions for ways to facilitate sustainability conversations and activities based on these two stories.

"The Ones Who Walk Away from Omelas" as an Allegory for Teaching Sustainability Topics

In its simplest sense, "The Ones Who Walk Away from Omelas" supplies a model for thinking about existing social structures, both regional and global, that perpetuate inequalities, and by extension, produce unnecessary suffering. Students will be quick to provide examples of inhumane exploitation from indigenous genocide, to sweat shops, fire, flood, and famine victimization, to garbage and sewage dumping, or other instances of environmental contamination. The costs of additional hidden or invisible (to some) violations of the social contract or public trust may be more difficult for students to identify and address. Despite the fantasy circumstances of Omelas, it is important to insist upon a literal reading of the imprisoned and neglected child's plight, certainly shared by a member of each student's own extended community. By reading this youthful character as a figure for an exploited factory worker, individual war casualty, climate refugee, colonized subject, or natural disaster survivor, students can better understand the intersecting demands of sustainability activists. United Nations SDGs #1 No Poverty, #2 Zero Hunger, #3 Good Health and Well-being, #10 Reduced Inequalities, and #11 Sustainable Cities and Communities can illuminate correspondences between Le Guin's narrative and real-world eventualities. This orientation opens students to conversations on the ethics of care, basic postcolonial theory, and utilitarian philosophy or pragmatism by which the good of the many outweighs the good of the one, as in the community of Omelas.[38] Students tend to be fascinated by discussions of pragmatic Utilitarian philosophy or "trolleyology" (figure 8.1).

As the simpler of the two stories, "The Ones Who Walk Away from

Figure 8.1 "File:Trolley Problem.png." *Source*: By McGeddon. McGeddon is licensed under CC BY-SA 4.0.

Omelas" can be used to introduce concepts and themes that will be further developed in "The Seventh Event." Beyond the dilemma of allowing an innocent to suffer to ensure the well-being of the society, Le Guin introduces a variety of compelling themes that students can define and relate to shared realities and their own experience. The timeless perfection of a utopia or perfect world, for example, depends upon stasis. This rigidity stands in opposition to solution-seeking and productive change. When students determine that Omelas is, in fact, a flawed, or imperfect, place, even a dystopia, their evidence will prompt further inquiry into the facts and causes of social inequality. The plight of the one against the many is balanced with the individual Omalasians' decisions to stay versus leave. None takes a collective action in this regard. Thus, individual accountability is set against a somewhat dehumanized view of society as an intersectional system. Or has Le Guin, in fact, hit upon the truth of human selfishness, of extractive consumer society? Students may look at the suffering child as an allegory of poverty and critique abundance versus scarcity models in media representations of the world and its people. The artistry of Le Guin's opening paragraph contrasts with the brutal logic both narrator and the Omalasians use to justify the child's imprisonment. Both realms of creative thinking, the artistic and the scientific, will merge even more intricately in Richard Powers's "The Seventh Event."

"The Seventh Event" as a Literary Model of Systems Thinking for Sustainability

In Richard Powers's challenging story, the autobiographical narrator recounts conversations and other interactions with a lifelong friend Mia Erdmann in a way that may prompt students' consideration of their own histories and friendships. The narrator's life coincides closely with the author's; however, Mia is a completely fictional character. The story begins with her death and recounts, in reverse chronology, their discussions on environmental sustainability and theories of replicating life that tend to resonate with students. Mia first attempts to "make life" and thus preserve the planet as a biochemist, but later turns to writing fiction as a more useful way of reproducing reality. Powers presents intersecting environmentalist themes of species extinction and other threats to the planet linked to climate change and human impact through the remembered interactions of his characters. This text directly engages patterns of human behavior that threaten the sustainable future of the planet.

Powers puts poetry and science to the test as he explores the human capacity to imagine and understand the world, a fascinating topic for upper-level high school and university students. However successfully scientists construct technologies capable of measuring planetary features, including climate change and the rate of species extinction, these activities do not account for the scientific wonder that drives characters such as Mia Erdmann to remain curious and engaged, to reject her own death, as rationally explainable as it might be. Students may not be able to identify all the intertextual allusions, but they will be able to productively connect aspects of the story to the United Nations SDGs, especially themes relevant to #3 Good Health and Well-Being, #13 Climate Action, #14 Life Below Water, #15 Life on Land. For this story, The Earth Charter with its ethics-based call to action or Kate Raworth's intersectional Doughnut Model might be suitable frameworks to employ. Because this story engages the environmental crisis directly, it presents a platform for discussing specific categories of threat and destruction as they imperil certain demographic groups.

"The Seventh Event" involves many intertextual references, so an instructor could use it as a starting (or ending) point for a much broader module on types of nature writing. For example, this unit could be expanded to include discussions of the Romantic sublime's depiction of nature, including William Wordsworth's "Tintern Abbey," or nineteenth-century U.S. nature writing such as Henry David Thoreau's *Walden*, as well as contemporary versions of this tradition, including Annie Dillard's "Living like Weasels" from *Teaching a Stone to Talk* or Danez Smith's afrofuturism in *Don't Call Us Dead*. Powers's narrator refers to Mia as a "diabolical Annie Dillard." Students will appreciate examples of deep ecology with its celebration of the natural world as intrinsically good and valuable, such as Rachel Carson's *The Sea Around Us* or Gary Snyder's "Rip Rap." Then, there are modernist depictions of nature such as Jean Toomer's *Cane*; the paintings of Georgia O'Keefe; the literary naturalism of Jack London's "To Build a Fire" or "What Life Means to Me"; the feminist metaphors of Audre Lorde's "Coal," or the vivid explorations of Elizabeth Bishop's "The Fish"; indigenous traditions such as Leslie Silko's poem "The Deer Dance" or Robin Wall Kimmerer's *Braiding Sweetgrass*; or the social ecology of W. S. Merwin's "For a Coming Extinction" in which social/political contexts come into play.

Teaching sustainability topics through works of literature provides rewarding experiences for both students and educators. Fiction constitutes a dimension through which urgent and often terrifying challenges can be addressed with imagination and creativity to parallel the real-world solutions we need. Each story's or poem's implications change over time with rapid developments in pro-sustainability innovation. Students tend to find literary works more compelling when they understand direct implications in their own lives. Because "The Ones Who Walk Away from Omelas" and "The Seventh Event" (as well as other recommended works mentioned here) point to options and alternatives to continued inequality and suffering borne out by climate change, they become vehicles for inspiration and eventually, environmental, economic, and social change. Students will come away from these learning experiences with a strengthened sense of their own power to be part of that crucial transformation.

NOTES

1. Martha Nussbaum (2010). *Not for Profit: Why Democracy Needs the Humanities* (Princeton: Princeton University Press).
2. Ursula Le Guin. "The Ones Who Walk Away from Omelas." (1973) *The Wind's Twelve Quarters* (New York: Harpers, 1975): 275–84.
3. Richard Powers, "The Seventh Event" *Granta 90: Country Life* (Summer 2005): p. 57ff.
4. Nel Noddings, *Caring: A Feminine Approach to Ethics and Education* (Berkeley: University of California Press, 1984).
5. Neil Lewis, keynote address, "Sustainability and Resiliency in Times of Change," 2020 State of New York Sustainability Conference presented by the New York Coalition for Sustainability in Higher Education (NYSCHSE), 2 December, 2020.
6. Bret Eynon and Laura M. Gambino, *High-Impact ePortfolio Practice: A Catalyst for Student, Faculty, and Institutional Learning* (Stylus, 2017).
7. Lorin Anderson, et al. *Taxonomy for Learning, Teaching, and Assessing: A Revision of Bloom's Taxonomy of Learning Objectives* (New York, Longman, 2001).
8. Adapted with permission from Sarah Wyman, "Reading Through Fictions in Ursula Le Guin's *The Ones Who Walk Away from Omelas*." *ANQ: A Quarterly Journal of Short Articles, Notes and Reviews* 25, no. 4 (2012): 228–32.
9. Le Guin, *The Ones Who Walk Away from Omelas*, 188.
10. Le Guin, *The Ones Who Walk Away from Omelas*, 275.
11. Le Guin, *The Ones Who Walk Away from Omelas*, 278.

12. Le Guin, *The Ones Who Walk Away from Omelas*, 276.
13. Le Guin, *The Ones Who Walk Away from Omelas*, 277.
14. Le Guin, *The Ones Who Walk Away from Omelas*, 277–78.
15. Le Guin, *The Ones Who Walk Away from Omelas*, 278.
16. Le Guin, *The Ones Who Walk Away from Omelas*, 278.
17. Le Guin, *The Ones Who Walk Away from Omelas*, 283.
18. Ursula Heise, "The Hitchhiker's Guide to Ecocriticism" *PMLA* 121, no. 2 (2006): 503–16.
19. Heise, "The Hitchhiker's Guide to Ecocriticism," 511.
20. Powers, "The Seventh Event," 66.
21. Powers, "The Seventh Event," 62.
22. Powers, "The Seventh Event," 61.
23. Powers, "The Seventh Event," 59.
24. Powers, "The Seventh Event," 73.
25. Heise, "The Hitchhiker's Guide to Ecocriticism," 510.
26. Powers, "The Seventh Event," 61.
27. Powers, "The Seventh Event," 62.
28. Powers, "The Seventh Event," 64.
29. Powers, "The Seventh Event," 63.
30. Powers, "The Seventh Event," 71.
31. Powers, "The Seventh Event," 66.
32. Powers, "The Seventh Event," 72.
33. Powers, "The Seventh Event," 74.
34. Powers, "The Seventh Event," 72.
35. Powers, "The Seventh Event," 62.
36. Powers, "The Seventh Event," 64.
37. Powers, "The Seventh Event," 68.
38. Nel Noddings, *Caring: A Feminine Approach to Ethics and Education*. Berkeley: University of California Press, 1984.

REFERENCES

Akwasasne Notes, Ed.''Hau de no sau nee Address to the Western World" [1977] in *Basic Call to Consciousness*, Native Voices, 2005.

Anderson, Lorin et al. *Taxonomy for Learning, Teaching, and Assessing: A Revision of Bloom's Taxonomy of Learning Objectives*. New York: Longman, 2001.

Eynon, Bret and Laura M. Gambino. *High-Impact ePortfolio Practice: A Catalyst for Student, Faculty, and Institutional Learning*. Stylus, 2017.

Heise, Ursula. "The Hitchhiker's Guide to Ecocriticism" *PMLA* 121, no. 2 (2006): 503–16.

James, William, "The Moral Philosopher and the Moral Life." *The Will to Believe and Other Essays in Popular Philosophy*. New York: Dover, 1957.

Kimmerer, Robin Wall. *Braiding Sweetgrass: Indigenous Wisdom, Scientific Knowledge and the Teaching of Plants*. Milkweed, 2013.

Le Guin, Ursula. "The Ones Who Walk Away from Omelas." (1973) *The Wind's Twelve Quarters*. New York: Harpers, 1975: 275–84.

Lewis, Neil. "Sustainability and Resiliency in Times of Change," Keynote address, 2020 State of New York Sustainability Conference presented by the New York Coalition for Sustainability in Higher Education (NYSCHSE), 2 December, 2020.

Noddings, Nel. *Caring: A Feminine Approach to Ethics and Education*. Berkeley: University of California Press, 1984.

Nussbaum, Martha. (2010). *Not for Profit: Why Democracy Needs the Humanities*. Princeton: Princeton University Press.

Powers, Richard. "The Seventh Event" *Granta 90: Country Life* (Summer 2005): 57ff.

Smith, Danez. *Don't Call Us Dead*. Greywolf, 2017.

Sulitest, an initiative of the Higher Education Sustainability Initiative (HESI) and the United Nations Sustainable Development Goals partnerships. https://www.sulitest.org/en/test.html.

"Trolley Problem." Wikipedia. https://en.wikipedia.org/wiki/Trolley_problem.

Wyman, Sarah. "Reading Through Fictions in Ursula Le Guin's *The Ones Who Walk Away from Omelas*." *ANQ: A Quarterly Journal of Short Articles, Notes, and Reviews* 24, no. 5 (2012): 228–232.

Chapter 9

Developmental Bibliotherapy and Cli-Fi

Helping to Reframe Young People's Responses to Climate Change

Judith Wakeman

Chance favours only the prepared minds.—Louis Pasteur[1]

Few Australians were untouched by the fires that raged from September 2019 till January 2020 across every Australian state, crushing the capacity of our firefighters to protect us during a period now dubbed The Black Summer. News stories of survivors, the persistent smoke haze, and the reported deaths of billions of animals threatened to overwhelm our emotions. For those directly impacted, these fires turned historic communities and favorite holiday destinations into nightmares.

Despite decades of warnings from climate scientists and emergency personnel, we were unprepared and powerless—only the forces of nature itself had the capacity to halt the destruction. Climate change was no longer something that happened decades away; species extinction and ecosystem collapse unfolded as we watched.

Young people are the group most vulnerable to the immediate effects of natural disasters: their social networks are more likely to be disrupted earliest and for longest; their go-to adults at home, school, or other activities may not be contactable or might be occupied elsewhere; their personal and individual needs may be considered less urgent than those of younger siblings, the wider community, or immediate family concerns; their emotional resilience and personal strengths are still being refined and developed; they will not be as able to identify their own feelings, articulate their needs, and locate the help they require; they are more likely to be overwhelmed by grief, trauma, guilt, hopelessness, and frustration; their coping behaviors may be misinterpreted or misunderstood, and they are less likely to voluntarily talk about their emotional well-being.

Coinciding with the onset of the fires in September 2019, The Australian Institute's "Climate of the Nation" report[2] stated that, while the incidence of concerns about climate-related issues were increasing in the adult population, those aged between eighteen and thirty-four were far more concerned about climate change (83%). Even more concerning is that 25% of young Australians think the world will end before they get old, and 66% say that climate change makes them feel helpless and afraid. Ten percent of young people believe it too late to do anything about climate change. Many climate scientists agree.[3]

It is not difficult to understand why issues related to climate change are concerning for adolescents. They are more informed of the science than previous generations and they will be more likely to experience the effects of escalating climate in their lifetimes. They see that nothing is being done to protect their future, they see the willful ignorance of politicians who make decisions that cause climate change to accelerate, but they themselves have no political influence. They are acutely aware that they are likely to become the first "global generation of children [to] grow up in a world made far more dangerous and uncertain as a result of a changing climate and degraded environment."[4] UNICEF calls climate change "a direct threat to a child's ability to survive, grow, and thrive" and estimates that 90% of the burden of disease attributable to climate change is borne by children under the age of five. According to UNICEF, a child born today could be living in a world with an average temperature that is four degree Celsius warmer by their seventy-first birthday.

On Australian national TV, a young woman tearfully shares her conviction that motherhood in the near future is unsafe, unethical, and a bad choice for the planet, while a young teenage protester addressing her peers asks for a curriculum that includes Climate Science, Media Literacy, and Political Engagement.

The high school student thinks about climate change every day, she reads about how ecosystems are on the brink of collapse and listens in despair as her teachers and parents tell her that it's up to her generation to fix things. A year 10 student at a "Schools Strike for Climate" rally talks about her fears for the future, her traumas framed by past experiences, and then, blending pleas with demands, she calls on those in power to take action on climate change, care for the environment, "listen to the Science," and "tell the truth." The second grader is scared about the planet but says it feels good to be surrounded by so many people (at the School Strike for Climate) who care, since he sometimes feels as if nobody else is worried; his parents are proud that their child is aware, but concerned that he could become overwhelmed by predictions that seem to be growing ever more disturbing.

But these young people's concerns are dismissed; instead, adults tell them to study hard so they can fix the climate when they graduate or become elected. And a senior Australian politician says he is "not worried about what might happen in 30 years' time"[5] when asked about climate change.

Understandably, young people are either increasingly becoming extremely concerned about climate change or in denial.

TEACHING CLIMATE CHANGE WITHOUT CREATING DESPAIR OR ENTRENCHING DENIAL

Is climate change education appropriate for young adults (YAs) in 2021? Even though the likelihood is that climate change will reach crisis point in their lifetimes, is this knowledge too great a burden for the young people in our care? Are we expecting them to find solutions for problems they have no ability to solve? How do we raise a generation to look forward to the future with hope when all they see around them are messages of gloom?

But, can we honestly say that we care about our students if we do nothing to prepare them for life in a climate-changed world?

As educators, we negotiate a narrow pathway as we aim for responsible education on climate change without inflicting further trauma. We must be able to nurture our students' resilience, teach them how to take responsibility for their own well-being, and prepare them for an uncertain future without paralyzing them with anxiety. We must be open to discussion and able to advocate for scientific viewpoints, even in the face of our own lack of expertise or understanding, but calmly when confronted with denial. And we must be prepared to acknowledge, identify, and respond appropriately to our students' anxiety or distress.

Teachers, and parents, must not lose sight of this pathway between being honest and being reassuring, between empowering young people with hope and weighing them down with responsibility. Some adults might say that it is the rhetoric surrounding climate change that is creating anxiety in young people. But children and adolescents will register our concerns via overheard conversations, news items, popular films that generate questions among their peers, or by experiencing the effects of wild weather and natural disasters unfolding around them. Young people's anxiety is fueled by the inaction of adults even as it drives their activism.

Young people want a curriculum that prepares them for the uncertainties of a warming planet, with reliable information about the facts and the magnitude of the threat of climate change in honest, open, and frank discussions with trusted and informed adults, whether in the classroom or around the dinner table, in ways they can understand. Young people want their emotions and concerns acknowledged, they want to be able to make informed choices, and they want to feel empowered to make their own choices. They want to be able to access and influence policy makers, and they want to join with peers who think along the same lines.

ACTIVISM: WHERE HOPE RULES

For many, concerns about the threat of climate change manifest as Eco-Anxiety, defined as the chronic fear of environmental doom. Some also experience an emotion described as a "homesickness without leaving home" or a "loss of their sense of place" which Glenn Albrecht called Solastalgia.[6] Anxiety is a common emotion and can be a protective mechanism. Our evolutionary responses to anxiety of Fight, Flight, or Freeze are also symptoms of eco-anxiety. But eco-anxiety differs from most other forms of anxiety in that it is considered a rational, reasonable, and appropriate response to the very real and existential threat posed by climate change. Some people refuse to acknowledge either that climate change exists or that it is caused by human activity—this could be likened to the Flight response. Some people are overwhelmed, thinking that climate change is too distressing, too difficult to understand, or beyond their power to fix and refuse to discuss or even think about climate change—this could be considered the Freeze response.

For some the response to eco-anxiety manifests as Activism: this is the Fight response. In the context of climate change, the Fight response is considered to be the only adaptive and appropriate response to the existential threat posed by climate change; because, unlike an imagined threat, neither the freeze nor the flight response has any chance of eliminating the source of the anxiety.[7]

For many climate scientists eco-anxiety is akin to the despair felt by Cassandra, whose gift of prophesy was frustrated by the curse of not being believed; they watch as the science is ignored in favor of populist alternatives that exacerbate the accelerating ecological destruction and make their own scientific predictions even more likely. For them, action has become the antidote to despair. Richard Eckersley[8] identified three responses to fear of the apocalypse: Nihilism, Fundamentalism, and "Activism: Where Hope Rules." All anxiety contains an element of hope, and hope cannot coexist with despair. As hope inspires climate change action, and in turn, climate action generated by one's peers can generate hope, people transform their anxiety into action as part of a united social mass of like-minded individuals who want to see a brighter future.

Young people are already at the forefront of environmental activism. Their anxiety in the face of an uncertain future is appropriate, and taking action is a healthy and proven way toward addressing their concerns. These young activists may not be the teens about whom parents and teachers should be most concerned.

The Psychology of Denial

While the psychology of climate change denial is complex, it is worth remembering that

- climate Deniers are in the minority,
- most governments are taking action on climate change,
- denial is a maladaptive (unhealthy) response to climate change, and
- the most significant predictor of climate change denial is political affiliation.

Stanley Cohen distinguishes between denial that is personal and psychological and denial that is organized and collective.[9] He defines the first group of people as those who deny the facts to themselves for a variety of personal reasons. In the latter situation, facts are denied, manufactured, or misrepresented to others by institutions, even when truth is very well-known.[10]

Reasons for denial that is personal and psychological fall into these six broad categories:

- Fear, whether conscious or unconscious, may be a coping strategy related to self-preservation. We might find issues related to climate change are too upsetting, or make us feel guilty or ignorant.
- Self-efficacy and agency make us question our ability to adapt, achieve meaningful change, or even know who to trust.
- Identity—social, cultural, political, religious, and community identity—influences our personal responses to climate change. We want to fit in with our chosen group, and not be the one to rock the boat or be a "party pooper."
- Perception of risk, the need to juggle demands that compete for time or money, or conflict with our self-interest and social status, can determine what we are prepared to sacrifice at any point in time.
- Education, whether a knowledge deficit or information overload, our level of access and trust on this information, will determine our confidence in and willingness to act on that information.
- Proximity to the affects of climate change—how close are we to being personally affected either geographically or temporally—significantly affects our response.

Whatever the reasons, the consequences of denial are dire.

On Individuals:

- Denial can manifest as despair, numbness, apathy, nihilism, and fundamentalism.
- We feel hopeless, powerless, paralyzed, guilty, angry, and confused.
- We suffer loss of self-efficacy, self-sufficiency, self-reliance, and self-determination.

On Society:

- Decadence or Dogma rules.[11]
- Self-interest plays out in what has been termed the "Tragedy of the Commons" recognized as hoarding during pandemic lockdowns.

And on the Planet:

- Cascading tipping points;
- Runaway climate change;
- Social, economic, and environmental systems collapse.

Denial in the short term is a comfortable option, but denial in the face of the existential threat of climate change is thwarting the very actions needed to halt it.

HARNESSING THE POWER OF STORIES

Stories engage the emotions and reduce stress, opening the pathways to learning. Stories can give a broader audience a better sense of what is happening, framing responses from differing viewpoints, and making the science real and personal. Stories use metaphor and analogy to enable us to see ourselves and others from diverse perspectives, and to help us identify our goals and define our values. The ability to share stories across time and space distinguishes humans from other animals. Stories give us role models and fallible heroes, questions and answers, comfort, companionship, and ancient wisdom; it is no coincidence that great leaders read fiction.

Stories can serve a didactic function: we experience both vicarious learning and vicarious rewards as skilled authors speak through the voices of their characters, teaching us language and encouraging healthy patterns for behavioral, social, and emotional learning. Stories have the power to motivate and inspire us by nurturing self-efficacy, optimism, and resilience, which generate the hope and creative energy to act. Stories validate our experiences, encourage us to explore, reframe our problems, reflect us on our journey, and help us to accept our past and face our future.

Tell Me the Truth: Talking to Children about Climate Change

At what age do children start questioning adults' honesty? Research tells us that, partly because it is in their best interests to do so, children happily accept the existence of Santa Claus and the Tooth Fairy until the age of eight. But what effect does it have on children when they realize that the adults in their lives are not being honest with them? If we are not open and honest with young people we risk creating distrust, we risk alienating them, and we risk fostering anxiety, just as we also foster anxiety by overwhelming them with information they may not be ready to absorb.

And how do we support their activities and actions—some of which might have risky and far-reaching outcomes—if those actions serve to relieve their anxiety and distress? However we choose to talk about climate change with young people, eco-anxiety has been described by Dr. Sarah Anne Edwards[12] as an intelligent response to climate change that enables a better understanding and sensible choices and a movement from paralyzing emotions into empowering actions. In other words, a healthy, natural reaction to our growing consciousness of a real threat, one, she says, "we should not minimize, discount, distract or otherwise suggest palliatives to ease" because "the more society and those around us discount the reality of the consequences at hand, the more anxious we become and the more maladaptive our responses."

Giving people the opportunity to share their emotions related to eco-anxiety or solastasia demonstrates that someone cares to acknowledge that their feelings and concerns are real and valid. It also provides them with the language they need to identify and articulate how they are feeling. Participating in activities that direct their attention outward and encourage them to mix with like-minded people also fosters efficacy and hope, which can be important outlets to relieve eco-anxiety.

In the face of the ever-present indications of climate change in the news, in school curriculums, in reading matter, and in covert adult discussions, Dr. Blanche Verlie found that young people feel anxious, overwhelmed, guilty, frustrated, grief-struck, and powerless. But that hope is fostered when they engage with other people who share their feelings and concerns, and take action to address climate change.[13] According to many scientists, there's a huge amount we can still do to protect what's left and make a meaningful difference. Young people need to receive this message.

Enabling Conversations

Climate change has already robbed many young people around the globe of a future resembling one their parents enjoyed. Young people are aware that the changes needed to mitigate the effects of climate change are happening too slowly. Despite the catastrophic personal, economic, and environmental toll taken by Australia's "Black Summer" fires, government and media reports on bushfires, droughts, extinction crises, and economic forecasts rarely mention climate change. Young people, frustrated by political inaction on climate change, and watching as governments unite and mobilize to combat a global pandemic, could be forgiven for interpreting this response as "governments care about their elders, but they don't care about their youth."

It is clear that avoiding the topic of climate change is not making young people feel safe. In a 2019 *Washington Post* survey,[14] over 40% of teens believe their generation will be impacted by climate change, only 15% say their education includes "a lot" about climate change (possibly indicating that their education does not meet their desire or expectation), and yet the same survey found that only 10% of teens regularly discuss climate change with friends or family.

Conversations about climate change are difficult. Our perceived lack of knowledge, concerns that we might upset others, feelings that we can't really change anything, that there are more immediate problems, or that we might have to make too many sacrifices make it easier to keep silent.

Having someone acknowledge, name, and validate feelings of anxiety or grief is the first step toward regaining control over these feelings. But people don't like talking about things that make them feel unhappy or make others feel uncomfortable; instead they maintain a small group of like-minded people and are tentative when discussing climate change with strangers, or sometimes even with close friends.

It should be a major concern for educators that, despite 95% of young Australians believing that climate change is a serious problem, as in the United States, barely 10% of Australian adults and children regularly talk about climate change to friends or family.

Using Cli-Fi as a Conversation Starter

How can educators help young people generate the types of conversations they need and encourage them to ask the questions they want answered? One strategy is to use cli-fi[15] specifically written for YAs as a conversation starter. Cli-fi is a genre set in the present or the very near future with effects of climate change as a backdrop and a plot in which catastrophic events unfold amid social and environmental upheaval. YA cli-fi features teenage protagonists with absent or unhelpful adults and settings that remain on a local level.

The authors of YA cli-fi present readers with unsettling versions of future Earth, explore how families and relationships exist in changed societies, the nature of heroes and villains, and how we might coexist with an environment that has been callously disrupted.

Many cli-fi novels present readers with societies where interpersonal trust has disappeared, where institutions that have previously been able to help have collapsed, where animals and plants have acquired new and monstrous properties, and where nature and the weather are violent and unpredictable adversaries. In these worlds, where we have difficulty distinguishing good from bad, authors challenge us to reflect on what is important to us, what our values are, what we need to preserve, and what we are prepared to sacrifice.

Solar Punk author Sarena Ulibarri makes the observation that "any near-future science fiction that does not engage with climate change is fantasy."[16] Solar Punk is a version of cli-fi in which authors present better, fairer societies using the disruption caused by climate change to imagine creative and positive outcomes.

In their didactic role, authors of YA cli-fi explain scientific terms and concepts, explore individual and government responses, and explain other related concepts, providing their readers with knowledge and empowerment. Most importantly, they offer a positive perspective. In cli-fi we find heroes dealing with the impacts of climate change, adapting to the aftermath of natural disasters, and pressuring governments and corporations to act. Various and volatile combinations of fear, anxiety, confusion, and anger exist in YA cli-fi, but always with a message of hope.

In the classroom, cli-fi has the potential to raise awareness of issues related to climate change in a nonthreatening and nonconfrontational way, motivating debate and inspiring action, and so potentially alleviating anxiety. Cli-fi adheres to scientific accuracy, introducing concepts and language such as feedback loops, tipping points, permafrost, gyres, and gulf-streams—terms that beget curiosity, enable discussion, and encourage research and investigation.

Inside the pages of YA cli-fi, authors can discuss climate change facts and fallacies in simple terms, informing without lecturing, and challenging readers to reflect on their experiences and viewpoints without implying criticism or blame. By combining cli-fi with Developmental Bibliotherapy programs designed to stimulate classroom discussion that encourage self-reflection, young people can share their experiences and validate their concerns, explore and initiate opportunities for action, and consider the ways their lives might change, within the safe environment of the classroom supported by trusted librarians, teachers, and authors.

Readers become better equipped with the knowledge and the language to more confidently discuss climate change issues and concerns in other classes, in their friendship groups, within their families, and within community action groups. In this way cli-fi could play a significant role in helping young people identify their own eco-anxiety as a valid and reasonable emotion, take responsibility for managing their own well-being,[17] and provide opportunities for the action to combat despair.

Imagining the Future

But cli-fi has yet another purpose, and that is to show us our possible futures. Gregers Anderson identified five themes in cli-fi that challenge us to consider what these futures might look like. He called these themes The Social Breakdown, The Judgment, The Conspiracy, The Loss of Wilderness, and The Sphere.[18] When we read dystopian and science fiction novels, we explore other worlds with scant regard to how those worlds evolved. Cli-fi fills in those gaps, urging us to envisage the collapse of societal and ecological systems, extreme weather events, the morality of preppers, what martial law might look like, and even how a changed and changing world might impact on our health, forcing us to confront our grief and perhaps motivating us to fight harder to save those things we care about, to drive change, join rebellion, and embrace activism.

As we read cli-fi we might feel comforted by the realization that we can all be heroes and while we might not be ultimately responsible for saving the places we love, we must not walk away from the attempt.

READING AND WELL-BEING

The stories that were shared (because stories must be shared) around the earliest campfires would not have been so different to the stories we enjoy today. Stories would have related daily events, journeys, scandals, meetings, relationships, and battles, lavishly embellished to ensure they were firmly fixed in the listeners' memories. Stories would have been intended to inspire, to encourage, to teach, to motivate, to build character, to help individuals build a sense of identity and purpose within their community, and to help the community coalesce.

Reading stories helps to promote well-being in four ways:

- By Activating the Default Mode Network (DMN),
- By Cultivating Core Confidence,
- By Demystifying Mental Health,
- By Reframing Responses to Climate Change.

Activating the DMN

The DMN has been identified as a network of interconnected regions in our brains which is active during REM sleep, while involved in non-task related thought (daydreaming) and, almost paradoxically, while reading. It is in a constant performance/feedback/revision loop.

The DMN is responsible for the following:

- Forming our personal narrative: If we are aware of our personal traits, our learning styles, and our strengths and weaknesses, we can be more confident about choosing activities and learning new things. Self-reflection is an important part of this process.
- Developing a Theory of Mind: An awareness of our mental states and that the mental states of other people may be different to ours forms the basis of empathy and moral reasoning, and enables us to predict and interpret the behaviors of others.
- Filing, retrieving, and applying memories: We use our memories when we interpret current situations and when we are imagining the future.
- Vicarious learning (also called observational learning) is second only to mastery in how effectively we learn new skills.[19] As well as boosting our core confidence and emotional resilience, vicarious learning gives us the tools to learn and make predictions from simulations and the experiences of others.
- Logic and Patterns: Our ability to follow logical pathways and uncover patterns enables us to assimilate new ideas into our previous knowledge, and helps us to organize and learn from the myriad daily activities in our chaotic lives.
- Insight and intuition, creative solutions, the "Aha moments," lateral thinking, spontaneous thoughts, snap decisions, and spontaneous reactions are all associated with activity in the DMN. A sentinel feature within the DMN monitors the external world while our consciousness is captivated, and we respond "without thinking" when necessary, perhaps because that part of the brain responsible for determining the appropriate response is already engaged.
- Eudemonia: A study on the links between music and happiness has found "intriguing evidence [suggesting] that the DMN may help orchestrate both pleasant and eudemonic brain states."[20]

In short, the DMN is responsible for all the processes that make us who we are as individuals and as members of the society in which we find ourselves.

With the DMN responsible for all these activities it is clear to see why daydreaming and sleep are important in maintaining our well-being. Reading activates the DMN during the comprehension of a story in the formation and recall of the memories that enable us to follow the narrative. The DMN is engaged when we interpret and visualize a story's characters, events, and settings using memories of our past experiences. The DMN is engaged when we imagine ourselves in situations similar to those unfolding in the story we are reading and when we imagine how the characters will fare as the story progresses. The DMN is engaged when a story triggers reflections and memories of something in our past.

It is important to note that the DMN can be activated while exercising, driving, listening to music, or even playing with the dog. However, reading stories is the only activity for which activating the DMN is a necessary requirement.

Cultivating Core Confidence

Of the many personal attributes associated with well-being, Alexander Stajkovic[21] identified four as being essential to what he called Core Confidence:[22]

- Hope—having realistic goals and manageable pathways. Stories help us identify the dreams and values which form the basis of our goals and show us pathways toward those goals. With Hope we become flexible and adaptable, and we keep going when faced with adversity.
- Efficacy—our belief in our ability to create change. Efficacy determines our confidence and willingness to take risks and learn new things. With Efficacy we take responsibility for our own lives and for the lives of others. Efficacy is cultivated primarily through mastery but also significantly through vicarious experiences.
- Resilience—an ability to recover from setbacks. Resilience embraces the attributes of curiosity, mindfulness, self-reliance, empathy, and an ability to learn from failure. Resilient individuals are reflective and take responsibility for their own well-being. Stories about resilient individuals show us that Resilience can be learned and developed.
- Optimism—a realistic belief that we are responsible for our own happiness and that more good things will continue to happen to us in the future. Stories inhabited by role models who overcome negative influences and remain focused on their goals rejuvenate our optimism.

When we read about ordinary people achieving extraordinary successes or performing extraordinary deeds we come to realize that the brain is flexible and adaptable, that both intelligence and confidence can be developed, and we start to adopt a growth mindset:

> In a growth mindset students understand that their talents and abilities can be developed through effort, good teaching and persistence. They don't necessarily think everyone's the same or anyone can be Einstein, but they believe everyone can get smarter if they work at it.[23]

Stajkovic adds a word of caution, noting that the attributes of Core Confidence must be realistic and in balance; too much Hope anesthetizes us, keeping us passive when we should act; too much Efficacy breeds arrogance and we neglect our training, believe we have nothing left to learn, or reject new ideas and suggestions; too much Resilience makes us overly tolerant of adversity, or resigned and apathetic in the face of danger; and too much Optimism shrouds us in illusions and irrational beliefs, or causes us to waste energy on unattainable goals. But he also notes that these attributes are in a state of flux, so that too much is much better than too little.

Although Stajkovic was investigating well-being in the workplace, he noted that

> confidence [is] a human strength that may contribute to different forms of human betterment. By studying confidence, as a mechanism that psychologically enables one's potential, we may be better able to discover the best people have to offer, especially in today's turbulent world.[24]

Demystifying Mental Illness

Mental health, whether for themselves or for someone they care about, is a major concern for 70% of teens across all educational and socioeconomic groups, and only 4% of young people did not consider mental health issues to be of any concern at all according to a survey conducted in the United States.[25]

In Australian schools, one in seven teens will have met the criteria for having a probable serious mental illness in any given year.[26] The same research team reports that 25% of mental health issues originate during adolescence. Although the research by Kelly and colleagues was conducted to evaluate the Youth Mental Health First Aid program, it makes a strong case for the importance of early intervention in emerging mental health issues, and also points out that the barriers to early intervention are stigmas and stereotypes compounded by insufficient or incorrect information. But young people seek help reluctantly, and rarely, if ever, discuss mental health issues with peers; fear, confusion, and unfamiliar language, exacerbated by stigmas and stereotypes, almost certainly contribute to their silence.

The earliest portrayals of mental illness in YA literature were of characters struggling to come to terms with emerging mental health issues and leaving the narrative following some form of personal catastrophe that initiated medical intervention. More recently, YA literature portrays characters with greater agency in their mental health. Authors, often writing from personal experiences, portray protagonists who acknowledge their mental health issues; story lines unfold that see these characters, and those around them, accepting and managing this aspect of their lives with realistic and hopeful outcomes. This is an important advance in the way mental health issues are portrayed in YA fiction, especially since the authors take their responsibilities seriously to ensure their readers sense an honesty and authenticity, and come to trust the voices within these novels.

Speaking through the voices of their protagonists, trusted authors encourage empathy and compassion, promote efficacy, introduce appropriate language to promote discussion, address and deconstruct stigmas and stereotypes, replace fear with information, remove feelings of isolation, provide role models, validate experiences, create distance and perspective, provide simulations and prepare adolescents for possible futures, introduce us to the unreliable narrator and the fallible hero, and increase readers' awareness of developing mental health issues whether in themselves or people they care about.

Stories help young people learn how and when to ask for help—either for themselves or on behalf of friends or family members—reducing fear associated with stigmas and stereotypes so they are not afraid to step up to the task when called. In this way stories can help to enable early intervention, potentially mitigating the disruption caused by emerging mental health issues during adolescence.

Reframing Responses to Climate Change

A report published just prior to the summer of 2020 states that 95% of Australian youth believe that climate change is a serious problem, 80% are anxious about their future, and at least 15% are losing sleep due to worry about climate change. An astounding 75% of young people say their concerns are not being taken seriously.[27]

This same report states that "communication, modelling and action around climate change can make young people feel empowered and help them cope with distressing feelings about climate Change."[28] It recommends that education providers include more education on climate change (including causes, impacts, and solutions) in core subjects, and establish, recognize, and support pro-environmental activities.[29]

But while education is critical for informing and supporting children's understanding of climate change, young people are reporting that they do not have sufficient education regarding the climate crisis, nor about how individuals and communities can help address it.[30] As mentioned previously, some teachers may feel their own knowledge is insufficient for this task, or fear their concerns will inflame students' anxiety. On top of this, some parents might challenge educators who introduce climate change issues into the curriculum without formal consultation.

And yet, cli-fi can provide a useful bridge between studies in science, social justice, creative arts, critical thinking, literature, economics, politics, mathematics, and even law. Authors writing YA cli-fi explain terminology and concepts in ways relevant to and accessible by this age group, replacing ignorance with information, both asking questions and presenting answers in the language of their young readers. Through the eyes of the protagonists, readers recognize the urgency and magnitude of the threat from a safe distance, they learn to challenge myths and conspiracy theories, to unpack scientific metaphors and analogies, to recognize and confront bias and self-interest, and to validate their own concerns.

As cli-fi activates our DMN we are analyzing and evaluating new knowledge and information; considering and differentiating between alternative viewpoints and conflicting priorities; envisaging and collaborating in creative solutions; discovering different ways to contribute; reflecting on our values, goals, strengths, weaknesses; learning from others' experiences; discovering ancient wisdom; and acknowledging our grief. The actions of the DMN are cultivating our Core Confidence as we follow fictional characters on quests and through battles that empower us and reinforce our resolve, inspiring us to keep going despite our failings, leading us in the understanding that no one is perfect but that everyone has something to offer.

Supportive fiction, by definition, nurtures Hope, Resilience, Efficacy, and Optimism. But along the way it encourages curiosity, imagination, collaboration, empathy, laughter, relationships, values, pathways, perspectives, validation, insight, catharsis, knowledge, skills, universality, self-knowledge, self-acceptance, growth, healing, escapism, and vicarious learning. Good cli-fi can offer all of these because it shows us that life goes on, presenting us with opportunities to reframe and possibly rewrite our futures. In a cli-fi novel we learn that we are part of a bigger picture, that we can make a difference; we start to adopt a growth mindset and we find an antidote to despair in the actions, hope, and community of like-minded people.

Cli-fi helps us find something worth fighting for; we are validated and empowered, unified and supported, and part of a global movement, huge and historical, that has come together to make the world a better place. Most importantly, cli-fi encourages those in the unhealthy mindsets of Flight and Freeze responses to move into a growth mindset by cultivating hope and self-efficacy, and inspiring them to focus their energies outward. Educators must ensure that climate change education cultivates hope and cli-fi can do that by empowering young people to be agents for change.

Cli-fi can help us prepare for the unsettling and distressing realities of climate change, adjust to the realization that our environment is changing rapidly and unpredictably, identify the values and relationships most important to us, develop the emotional resilience to face the sadness and injustice accompanying climate change with courage and determination, and learn that adaptation is not just coping, not just resilience, not just transformation but also the capacity to form meaningful connections with others. We come to understand the passion behind G. K. Chesterton's assertion that "the true soldier fights not because he hates what is in front of him, but because he loves what is behind him."

DEVELOPMENTAL BIBLIOTHERAPY AND YA CLI-FI

In the 1970s Rhea Rubin defined Developmental Bibliotherapy as a branch of Bibliotherapy that "emphasizes clients' reactions and insights" and aims "to promote normal development and self-actualisation or maintain mental health."[31] She made a clear distinction between Developmental Bibliotherapy and therapeutic or clinical bibltiotherapy. Developmental Bibliotherapy can provide support and scaffolding for adolescents during a time now considered a "second 'window of opportunity' for [influencing] the development of childrens' brains and thus their futures."[32] In Developmental Bibliotherapy teachers and librarians develop activities that guide their students to make positive changes in their thoughts, feelings, actions, or beliefs; activities are designed to promote self-reflection, so that these young people are conscious of the changes taking place within themselves. Conducted informally in classroom or library settings, Developmental Bibliotherapy is not intended to be assessed; while the extent and style of the programs can vary[33] it is the process of reflection that is most important. The key elements of Developmental Bibliotherapy in 2021 can be simplified as follows:

- Students read YA fiction from trusted authors.
- Reflection is encouraged with activities conducted during class time, during guided discussion or casual conversations.
- Trusted adults, such as teacher librarians and classroom teachers, are responsible for directing the activities and discussion.

The twenty-first century aims of Developmental Bibliotherapy seek to

- provide authentic role models,
- offer insight and guidance,

- provide information and understanding,
- help teens identify and prepare for typical adolescent issues,
- promote personality growth,
- help teens thrive,
- promote and nurture well-being as a preventative strategy.

Trusted Authors of YA Fiction

Trends in award-winning YA literature show that young people are preferring "edgy" and realistic stories that address more confronting topics, including social issues and mental health, and prizewinning stories are becoming less didactic and more open-ended.[34] Still, while current YA fiction is becoming rich with diverse viewpoints, these are woven into familiar themes that include stories of epic journeys and quests, adventure, horror, crime, mystery, romance, and tragedy.

YAs demand authentic role models, so they are willing to empathize with fallible heroes and unreliable narrators. The protagonists in YA novels are often teens with autonomy and independence, while adults are either portrayed as trustworthy but remaining in the background acting as safety nets for reassurance or purveyors of essential information, exist as representatives of undesirable or fixed mindsets, or are physically or emotionally absent. As noted, the authors themselves are able to introduce valuable information and age-appropriate terminology to their readers without making them feel embarrassed or ignorant. It must be noted that YA authors often expose challenging truths; the school librarian, with their expert knowledge of the YA literature in the school collection and an awareness of students' reading patterns, has an important role in supporting student well-being and can work with teachers and students as they make reading selections.

YA literature allows teen readers to experience life vicariously, exploring possible scenarios and giving readers strategies with which to face their fears. In the characters of YA novels readers can find confirmation and validation of their concerns and emotions, and of their unique and valuable contribution to the world; they learn to see a wider view of normal and to feel normal themselves. It is the responsibility of the authors of YA fiction to always leave their reader with messages of hope for the future.

Encouraging Reflection

Humans are the story species says Joseph Gold in his book *Read for Your Life*.[35] It is the ability to share stories and pass them down through time and space that distinguishes humans from other animals. Storytellers and their readers know that stories are important for our own well-being, and by extension the well-being of others. The key element of Developmental Bibliotherapy is self-reflection arising as a reader identifies with significant characters or events in a story, which activates the DMN.

Educators can initiate this reflection and prompt discussion simply by asking probing questions.

Activities[36] could include book reviews, art activities, making bookmarks with favorite quotes, creating book posters, creative writing and fan fiction, debates, or drama activities. It is important that educators are prepared to assist their students in achieving closure after discussing controversial or challenging topics. This can be by way of suggesting a variety of possible solutions, or by diffusing tensions with a short fun exercise at the end of the lesson. Cli-fi also provides opportunities to invite teachers from other subject areas to add valuable insights to the discussion.

Librarians as Trusted Adults

Librarians have a special role to play in the school community. At a time when young people need trusted adults to guide them through fake news and conspiracy theories, librarians are in the top five professional groups the public believe are most likely to provide trustworthy information. But students are aware of school librarians' other unique qualities. Librarians are easy to find (they're almost always in the library) and almost every student has met the school librarian at least once. Librarians don't set tests and they don't make judgments.[37] Librarians are good at asking the right questions and listening to answers. Librarians love talking about books, especially YA literature, so they know about trigger warnings and can identify reading patterns. Many students see the school library as a sanctuary, and conversations about books often bring about a relationship with their school librarian different from their relationship with other teachers.[38] School librarians can also provide support for student well-being with displays of inspirational biographies, motivational quotes, bookmarks, and themed fiction displays.

Reading Provides Lifelong Benefits

Developmental Bibliotherapy programs that use YA fiction and encourage readers to reflect and engage with stories can encourage a growth mindset that empowers, enabling young readers to take responsibility for their own well-being into adulthood. That responsibility can take the form of an awareness of the need to seek help, but may also be an impetus to join a book group, start reading to children, or browse a bookshop during the lunch hour.

By overlooking and minimizing the importance of stories in the development of young people during all stages of their schooling, we risk failing in our responsibility as educators to foster their ability to effectively manage their own mental health and the lifelong skills they will need to face an increasingly uncertain and frightening future.

DEVELOPMENTAL BIBLIOTHERAPY, CLI-FI, AND DISSENT

There are many reasons why young people do not talk about climate change. But silence on climate change is a form of denial. It is a maladaptive (unhealthy) response to climate change. "Action is the antidote to despair" is a quote attributed to Joan Baez, and echoed in the fight to halt climate change across the globe. Cli-fi as a learning tool has the potential to motivate and inspire this necessary action.

O'Brien, Selboe, and Hayward identified three forms of action which they called Dutiful Dissent (effecting policy change by working through existing institutions), Disruptive Dissent (activism that seeks to challenge power relationships often through direct protests and collective organization), and Dangerous Dissent (generating new and alternative systems, relationships, and ways of organizing society).[39] They explain that

> in engaging with climate change young people are entering into debates that involve dissenting from prevailing norms, belief and practices . . . dissent [that is] closely linked to issues of social injustice, poverty and violence, as well as to environmental issues such as pollution and biodiversity loss.[40]

While these debates provide a valuable impetus for discussion and investigation, expression of any form of political dissent requires support, including education, to enable young people to reflect critically, and respond safely and appropriately.

Cli-fi alone will not save the planet from climate change, but in Developmental Bibliotherapy reading programs cli-fi can inspire young people to take an active role in government, can motivate them to take civil action, and can reveal alternative futures so that young people start to develop a growth mindset, and embrace the adaptive (healthy) fight response to climate change.

Ever since humans harnessed fire to extend their day beyond sundown, storytellers have shared stories—to explain the world, to calm our fears, to preserve ancient wisdom, and to guide the paths of future generations. Cli-fi authors are continuing that tradition.

NOTES

1. On December 7, 1854, at the opening of the new Faculté de Sciences at Lille, Louis Pasteur described the discovery of the link between electricity and magnetism as "by chance, you might say, but remember that in the fields of observation, chance favors only the prepared minds." "File:Louis Pasteur Université de Lille 1854-1857 dans les champs de l'observation le hasard ne favorise que les esprits préparés.pdf." *Wikimedia Commons, the Free Media Repository*. 22 Oct. 2020, 04:49 UTC. 24 Aug 2021, 02:01. https://commons.wikimedia.org/w/index.php?title=File:Louis_Pasteur_Universit%C3%A9_de_Lille_1854-1857_dans_les_champs_de_l%27observation_le_hasard_ne_favorise_que_les_esprits_pr%C3%A9par%C3%A9s.pdf&oldid=497095743.

2. Richie Merzian, Audrey Quicke, Ebony Bennett, Rod Campbell, and Tom Swann. 2019. "Climate of the Nation 2019 Tracking Australia's Attitudes towards Climate Change and Energy." The Australia Institute. australiainstitute.org.au/wp-content/uploads/2020/12/Climate-of-the-Nation-2019-WEB.pdf. 29.

3. Joe Duggan. n.d. "This Is How Scientists Feel." Is This How You Feel? Accessed June 25, 2021. www.isthishowyoufeel.com/this-is-how-scientists-feel.html.

4. United Nations International Children's Emergency Fund (UNICEF), 2016. "Environment and Climate Change." UNICEF. https://www.unicef.org/environment-and-climate-change.

5. Jade Macmillan. 2021. "Deputy Prime Minister Flags Excluding Agriculture from 2050 Climate Target." www.abc.net.au. ABC. February 7, 2021. https://www.abc.net.au/news/2021-02-07/michael-mccormack-flags-excluding-agriculture-climate-target/13130160.

6. Glenn Albrecht. 2012. "The Age of Solastalgia." The Conversation. August 7, 2012. https://theconversation.com/the-age-of-solastalgia-8337.

7. Action that engenders hope has been identified as a strong predictor of recovery from anxiety disorders.

8. Richard Eckersley. 2008. "'Nihilism, Fundamentalism, or Activism: Three Responses to Fears of the Apocalypse' by Eckersley, Richard—the Futurist, Vol. 42, Issue 1, January-February 2008 | Online Research Library: Questia." *The Futurist* 42 (1). https://www.richardeckersley.com.au/attachments/Futurist_Apocalypse_2008.pdf.

9. Stanley Cohen. 2015. *States of Denial: Knowing about Atrocities and Suffering*. Cambridge, UK: Polity; Malden, Ma.

10. Naomi Oreskes and Erik M. Conway document this comprehensively in *Merchants of Doubt*, originally published in 2010.

11. Eckersley, 2008.

12. Sarah Anne Edwards. 2008. "Eco-Anxiety: An Intelligent Response." Eco-Anxiety. March 20, 2008. eco-anxiety.blogspot.com/2008/03/by-sarah-anne-edwards-phd-lcsw-articles.html.

13. Blanche Verlie, 2019. "Bearing Worlds: Learning to Live-with Climate Change." *Environmental Education Research* 25 (5): 751–66. doi: 10.1080/13504622.2019.1637823.757.

14. Lunna Lopes and Cailey Muñana. 2019. "The Kaiser Family Foundation/ Washington Post Climate Change Survey—Main Findings—9349." KFF. September 16, 2019. https://www.kff.org/report-section/the-kaiser-family-foundation-washington-post-climate-change-survey-main-findings/.

15. First coined by Dan Bloom in 2008, Cli-Fi, short for Climate Fiction, refers to fiction that takes place in a climate-changed world.

16. Diego Arguedas Ortiz. 2019. "How Science Fiction Helps Readers Understand Climate Change." BBC. January 15, 2019. www.bbc.com/culture/article/20190110-how-science-fiction-helps-readers-understand-climate-change.

17. The link between language and well-being is beyond the scope of this chapter. For a comprehensive paper on the subject see Lindsay Oades and colleagues' "Wellbeing Literacy: A Capability Model for Wellbeing Science and Practice."

18. Gregers Andersen and Munk Carsten Hansen. 2014. "Fiction Prepares Us for a World Changed by Global Warming." University of Copenhagen. November 10, 2014. humanities.ku.dk/news/2014/fiction_prepares_us_for_a_world_changed_by_global_warming/.

19. Albert Bandura. 1978. "Self-Efficacy: Toward a Unifying Theory of Behavioural Change." *Advances in Behaviour Research and Therapy* 1 (4). doi: 10.1016/0146-6402(78)90002-4, 205–7.

20. Eloise A. Stark, Peter Vuust, and Morten L. Kringelbach. 2018. "Chapter 7 - Music, Dance, and Other Art Forms: New Insights into the Links between Hedonia (Pleasure) and Eudaimonia (Well-Being)." Edited by Julia F. Christensen and Antoni Gomila. ScienceDirect. Elsevier. January 1, 2018. www.sciencedirect.com/science/article/pii/S0079612318300190. 145.

21. Alexander D. Stajkovic. 2006. "Development of a Core Confidence-Higher Order Construct." *Journal of Applied Psychology* 91 (6): 1208–24. doi: 10.1037/0021-9010.91.6.1208. 1209

22. Other models on improving student performance and well-being exist, such as Seligman's Signature Strengths and the International Baccalaureate's Middle Years Programme. However, I believe the model proposed by Alexander Stajkovic is uncomplicated and intuitive. Core Confidence directly aligns with the character attributes needed to tackle climate change, as well as the activities associated with the DMN.

23. James Morehead. 2016. "Stanford University's Carol Dweck on the Growth Mindset and Education." OneDublin.org. April 18, 2016. onedublin.org/2012/06/19/stanford-universitys-carol-dweck-on-the-growth-mindset-and-education/.

24. Stajkovic 2006. 1221.

25. Juliana Menasce Horowitz and Nikki Graf. 2019. "Most U. S. Teens See Anxiety and Depression as a Major Problem among their Peers." *PEW Research Center*. www.pewresearch.org/social-trends/wp-content/uploads/sites/3/2019/02/Pew-Research-Center_Teens-report_full-2.pdf.

26. Claire Kelly, Johanna Mithen, Julie Fischer, Betty Kitchener, Anthony Jorm, Adrian Lowe, and Chris Scanlan. 2011. "Youth Mental Health First Aid: A Description of the Program and an Initial Evaluation." *International Journal of Mental Health Systems* 5 (1): 4. doi: 10.1186/1752-4458-5-4.

27. The Australian Psychological Society Limited. 2019. "Young People's Voices in the Climate Crisis." APS. https://psychweek.org.au/2019-archive/.

28. The Australian Psychological Society Limited 2019, 16.

29. The Australian Psychological Society Limited 2019, 17.

30. Lunna Lopes and Cailey Muñana. 2019. "The Kaiser Family Foundation/ Washington Post Climate Change Survey—Main Findings—9349." KFF. September 16, 2019. https://www.kff.org/report-section/the-kaiser-family-foundation-washington-post-climate-change-survey-main-findings/.

31. Rhea J. Rubin. 1979. "Uses of Bibliotherapy in Response to the 1970s." *Library Trends* 28 (2): 239–52.

32. Nikola Balvin and Prerna Banati (Editors). 2017. "The Adolescent Brain: A second Window of Opportunity a Compendium" UNICEF Office of Research. Florence, Italy. www.unicef-irc.org/publications/pdf/adolescent_brain_a_second_window_of_opportunity_a_compendium.pdf, 7.

33. And indeed this author hopes that for some educators and librarians this section will provide explicit validation of the importance of your existing programs.

34. Richards Bodart, Joni, Ashley N. Barrineau, and Mary L. Flamino. 2011. "The Printz Award Winners under a Microscope: Content Trends, Committee Challenges, and Teen Popularity." *The Journal of Research on Libraries and Young Adults.* Young Adult Library Services Association.www.yalsa.ala.org/jrlya/2011/02/the-printz-award-winners-under-a-microscope-content-trends-committee-challenges-and-teen-popularity/.

35. Joseph Gold. 2001. *Read for Your Life: Literature as a Life Support System.* Markham, ON: Fitzhenry and Whiteside.

36. Judith Wakeman. 2019. "Activities & Printables." Developmental Bibliotherapy. November 18, 2019. read4life.today/activities/.

37. American Library Association. 2017. "Professional Ethics." Tools, Publications & Resources. May 19, 2017. www.ala.org/tools/ethics.

38. A personal example is the young boy who asked me if I could find him a book about a boy who was being bullied.

39. An example of Dangerous Dissent occurred during Australia's Black Summer fires when the Animal Rescue Craft Guild facebook page grew to a global membership of over 225,000 in eight months, supporting wildlife carers with koala mits, marsupial pouches, bat wraps, bird nests, and possum boxes. Individuals, families, and community groups ameliorated their grief with action and mutual support. The page is still operating with over 210,000 members.

40. O'Brien, Selboe, and Hayward 2018.

REFERENCES

Albrecht, Glenn. 2012. "The Age of Solastalgia." The Conversation. August 7, 2012. https://theconversation.com/the-age-of-solastalgia-8337.

American Library Association. 2017. "Professional Ethics." Tools, Publications & Resources. May 19, 2017. http://www.ala.org/tools/ethics.

Andersen, Gregers, and Carsten Munk Hansen. 2014. "Fiction Prepares Us for a World Changed by Global Warming." University of Copenhagen. November 10, 2014. https://humanities.ku.dk/news/2014/fiction_prepares_us_for_a_world_changed_by_global_warming/.

Balvin, Nikola and Prerna Banati (Editors). 2017. "The Adolescent Brain: A second Window of Opportunity a Compendium" UNICEF Office of Research. Florence, Italy. https://www.unicef-irc.org/publications/pdf/adolescent_brain_a_second_window_of_opportunity_a_compendium.pdf.

Bandura, Albert. 1978. "Self-Efficacy: Toward a Unifying Theory of Behavioral Change." *Advances in Behaviour Research and Therapy* 1 (4). doi: 10.1016/0146-6402(78)90002-4, 205–7.

Cohen, Stanley. 2015. *States of Denial: Knowing about Atrocities and Suffering*. Cambridge, UK: Polit; Malden, Ma. 9.

Duggan, Joe. n.d. "This Is How Scientists Feel. Is This How You Feel?" Accessed June 25, 2021. https://www.isthishowyoufeel.com/this-is-how-scientists-feel.html.

Eckersley, Richard. 2008. "Nihilism, Fundamentalism, or Activism: Three Responses to Fears of the Apocalypse." *The Futurist* 42 (1). https://www.richardeckersley.com.au/attachments/Futurist_Apocalypse_2008.pdf.

Edwards, Sarah Anne. 2008. "Eco-Anxiety: An Intelligent Response." *Eco-Anxiety*. March 20, 2008. http://eco-anxiety.blogspot.com/2008/03/by-sarah-anne-edwards-phd-lcsw-articles.html.

"File:Louis Pasteur Université de Lille 1854-1857 dans les champs de l'observation le hasard ne favorise que les esprits préparés.pdf." *Wikimedia Commons, the free media repository*. 22 Oct 2020, 04:49 UTC. 24 Aug 2021, 02:01. https://commons.wikimedia.org/w/index.php?title=File:Louis_Pasteur_Universit%C3%A9_de_Lille_1854-1857_dans_les_champs_de_l%27observation_le_hasard_ne_favorise_que_les_esprits_pr%C3%A9par%C3%A9s.pdf&oldid=497095743

Horowitz, Juliana Menasce and Nikki Graf. 2019. "Most U.S. Teens See Anxiety and Depression as a Major Problem among their Peers." *PEW Research Center*. https://www.pewresearch.org/social-trends/wp-content/uploads/sites/3/2019/02/Pew-Research-Center_Teens-report_full-2.pdf.

Kelly, Claire, Johanna M Mithen, Julie A Fischer, Betty A Kitchener, Anthony F Jorm, Adrian Lowe, and Chris Scanlan. 2011. "Youth Mental Health First Aid: A Description of the Program and an Initial Evaluation." *International Journal of Mental Health Systems* 5 (1): 4. doi: 10.1186/1752-4458-5-4

Lopes, Lunna and Cailey Muñana. 2019. "The Kaiser Family Foundation/Washington Post Climate Change Survey—Main Findings—9349." KFF. September 16, 2019. https://www.kff.org/report-section/the-kaiser-family-foundation-washington-post-climate-change-survey-main-findings/.

Macmillan, Jade. 2021. "Deputy Prime Minister Flags Excluding Agriculture from 2050 Climate Target." ABC. February 7, 2021. https://www.abc.net.au/news/2021-02-07/michael-mccormack-flags-excluding-agriculture-climate-target/13130160.

Merzian, Richie, Audrey Quicke, Ebony Bennett, Rod Campbell, and Tom Swann. 2019. "Climate of the Nation 2019 Tracking Australia's Attitudes towards Climate Change and Energy." The Australia Institute. https://australiainstitute.org.au/wp-content/uploads/2020/12/Climate-of-the-Nation-2019-WEB.pdf.

Mission Australia, and Black Dog Institute. 2017. "Youth Mental Health Report: Youth Survey 2012-16." Sydney, NSW: Mission Australia. https://blackdoginstitute.org.au/docs/default-source/research/evidence-and-policy-section/2017-youth-mental-health-report_mission-australia-and-black-dog-institute.pdf.

Morehead, James. 2016. "Stanford University's Carol Dweck on the Growth Mindset and Education." OneDublin.org. April 18, 2016. https://onedublin.org/2012/06/19/stanford-universitys-carol-dweck-on-the-growth-mindset-and-education/.

O'Brien, Karen, Elin Selboe, and Bronwyn M. Hayward. 2018. "Exploring Youth Activism on Climate Change: Dutiful, Disruptive, and Dangerous Dissent." *Ecology and Society* 23 (3). doi: 10.5751/es-10287-230342.

Oades, Lindsay G., Aaron Jarden, Hanchao Hou, Corina Ozturk, Paige Williams, Gavin R. Slemp, and Lanxi Huang. 2021. "Wellbeing Literacy: A Capability Model for Wellbeing Science and Practice." *International Journal of Environmental Research and Public Health* 18 (2): 719. doi: 10.3390/ijerph18020719.

Ortiz, Diego Arguedas. 2019. "How Science Fiction Helps Readers Understand Climate Change." BBC. January 15, 2019. https://www.bbc.com/culture/article/20190110-how-science-fiction-helps-readers-understand-climate-change.

Richards Bodart, Joni, Ashley N. Barrineau, and Mary L. Flamino. 2011. "The Printz Award Winners under a Microscope: Content Trends, Committee Challenges, and Teen Popularity." *The Journal of Research on Libraries and Young Adults*. Young Adult Library Services Association. http://www.yalsa.ala.org/jrlya/2011/02/the-printz-award-winners-under-a-microscope-content-trends-committee-challenges-and-teen-popularity/.

Rubin, Rhea J. 1979. "Uses of Bibliotherapy in Response to the 1970s." *Library Trends* 28 (2): 239–52.

Stajkovic, Alexander D. 2006. "Development of a Core Confidence-Higher Order Construct." *Journal of Applied Psychology* 91 (6): 1208–24. doi: 10.1037/0021-9010.91.6.1208.

Stark, Eloise A., Peter Vuust, and Morten L. Kringelbach. 2018. "Chapter 7—Music, Dance, and Other Art Forms: New Insights into the Links between Hedonia (Pleasure) and Eudaimonia (Well-Being)." Edited by Julia F. Christensen and Antoni Gomila. *Elsevier*. January 1, 2018. https://www.sciencedirect.com/science/article/pii/S0079612318300190.

The Australian Psychological Society Limited. 2019. "Young People's Voices in the Climate Crisis." APS https://psychweek.org.au/2019-archive/.

United Nations International Children's Emergency Fund (UNICEF). 2016. "Environment and Climate Change." UNICEF. https://www.unicef.org/environment-and-climate-change.

Verlie, Blanche. 2019. "Bearing Worlds: Learning to Live-with Climate Change." *Environmental Education Research* 25 (5): 751–66. doi: 10.1080/13504622.2019.1637823.

Wakeman, Judith. 2019. "Activities & Printables." Developmental Bibliotherapy. November 18, 2019. read4life.today/activities/.

Afterword

I began writing this afterword on a chilly July morning in mid-coast Maine. When I first came to Maine as a child, hot days were rare and air conditioners unheard of. A dip in the lake was sufficient to cool off on hotter days. We kept the shades down in the day and screens in the open windows at night. All that has changed with the rising temperatures, and many houses have at least one air-conditioned space. The privileged (among whom I am aware I number) still chase the cooler summer days north of Portland, and it's easy to remember how it used to be here. A lot more imaginative extension is required to reckon with global climate change. A slightly diminished escape from the heat pales compared to the disastrous, life-endangering conditions endured by vulnerable populations of people who are stuck in treeless hot zones in low-income neighborhoods in the United States.

I draw on memory of experiences in big cities, in the humid south, even in heat waves in New England, and claim that I know just how they feel. But do I really? Empathy has limits.

Pull the empathetic circle out a little wider, to encompass people living in coastal plains or on barrier islands—not just vacationers on the Outer Banks, but people in Bangladesh, say, or on a small island that is about to be swallowed by the rising Pacific. The automatic fellow-feeling that undergirds empathic responses with recognition of their feelings and perspectives gets harder the further apart we are. Empathy infamously weakens with distance, dissimilarity, and unfamiliarity.

Now turn our thoughts to the nonhuman animals. We are pretty good at feeling bad about a threatened whale or polar bear. But can we cross the species barrier to care enough about the fates of the uncharismatic creatures whose habitats we have ravaged, depleted, or altered beyond recognition? I can raise a flicker of feeling for the snapping turtle spy-hopping at the swimmers in the lake, because she's always been lurking out there near Turtle Rock. It's harder to imagine abundance that I never witnessed, but I believe the science that documents the decline in populations of species. I feel only a glum acknowledgment that many creatures are already gone, but it's difficult to miss those you've never known.

As for insects and plants, it's harder still to evoke empathy for their challenges, though I'm old enough to have met new antagonists, such as the opportunistic tiger mosquitoes, reproducing in the amount of water you can hold in a nickel and spreading ever northward. I have seen the bloom times shift, and I've engaged in annual exercises in futility, ripping out invasive plants. Like recycling, buying local food, providing my own grocery bags, and washing with cold water, I've done my bit out of a sense of duty.

But does empathy for the environment really move me to action? I still travel in cars and planes, indulging in what science fiction writer Gwyneth Jones labels "private transport hypocrisy." Every day of the year I flush my toilet with clean drinking water. I know that the generations to come will be horrified by much of what we in the developed world have done and continue to do, without thinking. How much can fiction reading do to raise our consciousness and provoke us to change our settled ways? And does an empathetic response to fiction translate to authentic concern for the real world? I have vibrated with intensities of shared emotion and perspective more frequently for nonexistent beings, the inhabitants of storyworlds, than I have ever felt authentic *Einfühlung* for the inanimate world that sustains our existence.

The human capacity for empathy, piggybacking on biological substrates that support human intersubjectivity, doesn't automatically extend beyond our immediate tribe to distant or dissimilar others or to the inanimate world. Yet skillful creators, who use their own empathetic imaginations to call invented beings out of thin air, can evoke authentic sensations of recognition, compassion, indignation at injustice, and even empathy for insensate elements (such as a crumbling cliff, eaten away by a rising ocean).

Much depends on the readers invited into imaginative co-creation by writers, and, I will suggest, on the teachers who construct reading circumstances for maximum impact. I have spilt a lot of ink arguing that there's nothing automatic linking an experience of narrative empathy and altruistic action taken in the real world. Indeed, I have argued that fictional texts receive strong empathetic collaboration from readers in part *because* these stories don't demand anything in return. A reader can safely indulge in shared feelings with a fictional character, secure in the knowledge that the character can't turn around and ask for help. So the goal of using literary works, in this case a growing corpus of superb young adult writing, to stimulate concern for, awareness of, and action on behalf of the planet faces a fundamental obstacle. The books won't do it alone.

Happily, the world still has teachers, professors, and librarians, magnificently represented by the ingenious experts whose chapters you have just read. Whether their strategies involve supplying nonfictional accounts and studies to bolster the messages of fiction, inviting students to engage in role-taking or perspective-taking exercises in response to their reading, or urging students to attend to nature when they create their own haiku, these teachers take the extra steps that make the results of reading more likely. With multiple aims in mind—of encouraging caring for the natural world, developing environmental awareness, and stimulating climate activism—these teachers also attune their students' literary sensibilities and creative expression to the needs of the world they live in. They do so with hope and encouragement, working to develop their students' growth mindsets and countering the despairing nihilism of some environmental dystopic writings.

Reflective discussion of literary texts, writing in response to creative prompts, and participation in bibliotherapeutic groups offer methods for situating a private reading experience in a social context that has strong potential to increase readers' resolve to take action on behalf of the planet. However, as all teachers already know from their experience in the classroom, we should be careful not to oversimplify the signs of success in reaching and engaging teen readers. What would engagement look like? We might witness changes in students' behavior, such as making new dietary resolutions, planting trees or gardens, participating in climate activism, or discovering an environmental vocation. But the signs of engagement might also be far more subtle, resulting in altered feelings toward others, awakened appetites for information, aroused attentiveness to their surroundings, art-making, and introspection. Just because these more private, interior responses don't show so readily on the outside doesn't mean that they are insignificant. Among other possibilities, teen readers' interior responses may prime them for later, more overt responses to calls for engagement and action. Every teacher who has been in the classroom for more than a decade knows that the impact of one's teaching often registers with students long after they have graduated.

What can teachers do to prepare students for adulthood and responsible citizenship, and how can literary reading play a role? The chapters in this volume are replete with thoughtful strategies and nuanced consideration of the challenges faced by committed teachers. I agree with many of the authors who emphasize the importance of hopefulness and individual agency, especially in the face of what many perceive as a climate emergency. If empathy can be channeled into optimistic intervention or even a predisposition to act, rather than lingering in a shared sense of horror or despair, the prospect of engagement that translates into action may improve.

A major obstacle to altruistic action, prosocial intervention, and helping responses (even among those who have been awakened to the dangers of a warming planet, pollution, species loss, and so forth) lies in the psychological phenomenon known as "diffusion of responsibility." In a large crowd with many witnesses, for example, many will hang back from helping a victim under the belief that somebody else must be taking care of it. Environmental concerns suffer greatly from diffusion of responsibility. Reflective discussion and writing can address the understandable, aversive responses that push responsibility away from the self toward distant authorities, multinational corporations, or governments. Cultivation of the individual's sympathetic imagination could make a difference.

In teaching about climate and encouraging empathetic responsiveness to different facets of the current and future crisis, good teachers will not paper over the conflicts of interest that emerge when empathy highlights competing subjects of concern. For example, the clean energy industry does not inevitably or even often employ the same workers who currently labor in fossil fuel extraction and the manufacturing jobs linked to burning nonrenewable fuels. James Meeks, writing in *The London Review of Books* (15 July 2021), puts the spotlight on displaced workers, remarking on the unsettling spectacle, when "an inspiring, utopian, once internationalist movement to save humanity from climate emergency comes across a once inspiring, once utopian, once internationalist movement to save humanity from capitalist exploitation, and walks on by" ("Who holds the welding rod?" 17). Leading students in conversations that invite empathy for different subject positions, including current and future generations, cultivates critical thinking and the capacity to make ethical judgments.

So even if imaginative literature by itself can't promise wholesale change in a generation's attitudes, the evocation of empathy for the inanimate world and its inhabitants can be enhanced into maturing consciousness by using and refining the creative strategies described in this volume.

Suzanne Keen
Clinton, New York

Index

About the Editor and Contributors

Anna Bernstein is a fifth-year middle school English teacher in Nashville who holds a bachelor's degree in English from the University of Florida and a master's degree in secondary English education from Vanderbilt University. Her work focuses on literacy practices that empower young people to see reading and writing as acts of joyful emancipation that have the capacity to make change in their communities.

María Encarnación Carrillo-García (Murcia University) holds a PhD in didactics of language and literature. She has worked as a teacher in all levels of the Spanish educative system, from infant education to tertiary education. Her research area of interest includes children's, Young Adult (YA), and adult literature as educational resources for the teaching of languages. She has been an active member of the Spanish research group of the University of Murcia, Didactics of Language and Literature, since 2010, and of the Spanish innovation group of the University of Murcia, Deucalión, since 2016, which is responsible for designing innovation activities for the teaching of language and literature. She is the author of *Roald Dahl and Children's and Young Adults' Literature in Language Teaching: A Proposal for Improving Literary Competence in the Classroom* (Digitum, 2011). Her more recent publications include "Iberian Geosophy in the Work of Carolina Coronado from a Didactic Perspective" in Mark Gant and Susana Rocha Relvas (eds.) *Transcultural Spaces and Identities in Iberian Studies* (Cambridge Scholar, 2020); "Children's and Young Adult's Environmental Literature and Environmental Education. Contributions from the Didactics of Language and Literature," in M. Campos & M. C. Quiles (eds.) *Repensando la Didáctica de la Lengua y la Literatura. Pradigmas y líneas emergentes de investigación* (Visor, 2019); and "Use of Literature, Resources and Innovative Methodologies in the English

Classroom" in *Portalinguarum* 32, 2019. ORCID: https://orcid.org/0000-0002-8741-8412. RESEARCHGATE: https://www.researchgate.net/profile/Maria_Carrillo-Garcia

Rachel Cohen is a professional technical/freelance writer and researcher based in New York. Her research is focused on environmental engineering, with specific emphasis on natural resource management and technoecology.

Timothy J. Duggan is a professor of English education at Northeastern Illinois University in Chicago. He received his BA from the University of California at Santa Barbara, his MA from the University of Nebraska, and his EdD from the University of South Dakota. He frequently conducts teacher workshops at the Chicago Shakespeare Theater. Tim has published teaching guides for *Hamlet*, *Julius Caesar*, and *Lord of the Flies* through Prufrock Press, and has published two collections of songs from Shakespeare and other well-known literature, as well as numerous book chapters and journal articles.

Lorraine Kerslake (Alicante University/GIECO-Franklin UAH) holds a PhD in children's literature and ecocriticism and is a specialist in language teaching. She teaches at Alicante University, Spain, and has worked as a translator of literary criticism, poetry, and art and published widely on children's literature and ecocriticism. Her current research areas of interest include children's literature, the representation of animals and nature in literature and art, and ecofeminism. She has been an active member of the Spanish research group on ecocriticism, GIECO, since 2010 and is also a member of the Research Institute for Gender Studies at Alicante University. Since 2016 she has been managing editor of the journal of ecocriticism Ecozon@ (*European Journal of Literature, Culture and Environment*): http://www.ecozona.eu/. She is author of *The Voice of Nature in Ted Hughes's Writing for Children* (Routledge, 2018). Her most recent publications include "Of Mice, Rabbits and Other Companion Species in Beatrix Potter's More than Human World" in *Reading Cats and Dogs: Companion Animals in World Literature*. Slovic, Scott, Françoise Besson, Zélia Bora, and Marianne Marroum (eds). (Lexington Books, 2020); "Revisiting the Myth of Pan: From Neverland to Narnia" in *Envisioning Change: Environmental Humanities*. Rivero-Vadillo, Alejandro and Flys-Junquera, Carmen (eds). (Vernon Press, 2020); "Hughes's Collaboration with Artists" in Gifford, Terry (ed). *Ted Hughes in Context* (Cambridge, 2018); "Ted Hughes: The Importance of Fostering Creative Writing as Environmental Education." *Children's Literature in Education: An International Quarterly*. Springer, Oct 2020; https://doi.org/10.1007/s10583-020-09427-4 ORCID: https://orcid.org/0000-0002-8104-2418.

Marek C. Oziewicz is professor of literacy education at the University of Minnesota, Twin Cities, and holds the Sidney and Marguerite Henry Chair in children's and YA literature at the College of Education and Human Development. He teaches courses on how literature empowers young people to develop a multicultural mindset, environmental awareness, and justice literacy. His publications include *Justice in Young Adult Speculative Fiction* (2015), which examines how nonfactual narratives help young people develop an understanding of real-life justice issues in the modern world; special issue of *The Lion and the Unicorn* on "Children's Literature and Climate Change" (2021), and a collection *Fantasy and Myth in the Anthropocene: Imagining Futures and Dreaming Hope in Literature and Media* (forthcoming 2022), which argues that the urgencies of the Anthropocene are primarily challenges to our story systems. Dr. Oziewicz's current research is on representations of climate change in children's literature, on the development of climate change literacy, and on anticipatory imagination in speculative fiction.

David Robinson is the head of the Department of Education and Curriculum Studies and a senior lecturer in the Faculty of Education at the University of Johannesburg, South Africa. He graduated with a PhD from the University of the Witwatersrand. He has presented papers at conferences in South Africa and internationally and has published academic articles in the field of literature and English studies. He has an interest in the teaching of literature, and within this field he has an interest in ecocriticism and social justice. ORCID: 0000-0003-0178-3249.

Ryan Skardal is an English teacher who has taught in Hong Kong, New Jersey, and now at Brentwood College School, which is located on Vancouver Island, British Columbia. He has previously contributed to *Developing Contemporary Literacies through Sports* and to the *New Jersey English Journal.*

Kaela Sweeney teaches English at Clarke Central High School in Athens, Georgia. She received her MEd from Vanderbilt University. Her work alongside her students focuses on using storytelling as a tool to create awareness and change in communities.

Natalie Valentín-Espiet is a graduate student in the Master of Arts in Teaching program at Northeastern Illinois University in Chicago. She received her BA in English from NEIU in 2019. Natalie's essay writing has received NEIU's English Department Methods and Practices Award, and her poetry has been published in the *SEEDS Literary & Visual Arts Journal.* This is her first book chapter publication.

Suhasini Vincent defended her doctoral thesis on Experimental Writing within the Postcolonial Framework of Indian Writing in English through a joint-supervision program between the University of Paris 3—Sorbonne Nouvelle—and the University of Madras in 2006. In 2009, she was awarded the second prize for the best defended thesis on topics relating to the Commonwealth, organized under the aegis of the SEPC (Société d'études des Pays du Commonwealth). Her research focusses on the legal scope of environmental laws in postcolonial countries and explores ecocritical activism in the essays and literary works of Arundhati Roy, Amitav Ghosh, and Yann Martel. She is at present an associate professor at the University of Paris 2—Panthéon Assas since 2007.

Judith Wakeman graduated as a teacher librarian in 2004; before that Judith taught Mathematics and Information Technology. Along the way Judith has studied earth sciences (including meteorology and climatology), botany and zoology, philosophy (particularly taoism), psychology and youth mental health first aid. Judith's ongoing investigation into Developmental Bibliotherapy is an amalgam of far-reaching global research into the many ways that school reading programs can help young people to manage their own well-being. She also continues to read widely from the YA and cli-fi genres. But in the Australian summer of 2019 Judith became aware of the widespread incidence of eco-anxiety in young people as wildfires raged across every Australian state, and she refocused her research to demonstrate how reading programs using cli-fi can help address eco-anxiety in YAs and why such programs should be considered a vital aspect of their education. Judith describes herself as an independent scholar and, as the global pandemic overshadows ongoing environmental upheavals, she shares her analysis of existing research both as a way to help young people, like her own grandchildren, develop and maintain their own well-being in this period of uncertainty, and as an antidote to her own eco-anxiety. Judith lives on the land of the Wurundjeri, Woi-wurrung people of the Kulin nation in Victoria, Australia.

Sarah Wyman, PhD, teaches twentieth and twenty-first century comparative literature with a focus on poetry, literary theory, sustainability, and the visual arts at the State University of New York at New Paltz, United States. She also directs the Faculty Development Center and is a co-coordinator of the SUNY New Paltz Sustainability Learning Community. She received her PhD from UNC Chapel Hill, her MA in creative writing from Hollins, and her BA from Brown University. Recent publications on verbal/visual intersections appeared in *Feminist Formations*, *ANQ*, *African American Review*, *Theatre*

History Studies, and *Journal of Learning through the Arts*. Website: https://faculty.newpaltz.edu/sarahwyman/.

Rebecca L. Young serves as a content specialist for the nonprofit education organization Cognia and as chief examiner for the International Baccalaureate's Middle Years Programme in Language and Literature. A former high school English teacher, her PhD work at SUNY Binghamton explored using literature as a lens for climate literacy. This is the focus of her first book, *Confronting Climate Crises through Education: Reading Our Way Forward*, and of her current research on how interdisciplinary instruction and assessment can prepare young people for the challenges of a changing world. She has contributed to *Critical Insights: Americans in Exile* and *Teaching Climate Change to Adolescents: Reading, Writing, and Making a Difference*. Her forthcoming (2022) publication offers a collection of chapters focused on the ways alternative forms of storytelling can engage K–12 students as ecologically conscious, globally minded problem-solvers in the climate crisis.

www.ingramcontent.com/pod-product-compliance
Lightning Source LLC
Chambersburg PA
CBHW070632100726
47907CB00007B/1954

* 9 7 8 1 4 9 8 5 9 4 1 3 4 *